Concepts of R
in Physical

D0822144

brary
ollege
onncoll.edu

Concepts of Reduction
in
Physical Science

Marshall Spector

Philosophical Monographs
Second Annual Series

Temple University Press
Philadelphia

Library of Congress Cataloging in Publication Data

Spector, Marshall, 1936-
   Concepts of reduction in physical science.

   (Philosophical monographs)
   Includes index.
    1. Science—Theory reduction.   I. Title.
II. Series: Philosophical monographs (Philadelphia,
1978-   )
Q175.S675   500.2′01   78-5441
ISBN 0-87722-131-6
ISBN 0-87722-127-8 pbk.

Temple University Press, Philadelphia 19122
© 1978 by Temple University. All rights reserved
Published 1978
Printed in the United States of America

ISSN 0363-8243

Contents

Acknowledgments

Some of the research for this book was conducted under a grant from the National Science Foundation, as well as two summer grants from the Research Foundation of the State University of New York.

The first four chapters are in part a much expanded and altered version of a paper entitled "Russell's Maxim and Reduction as Replacement" which appeared in *Synthese*, Volume 32, Nos. 1/2, Nov.-Dec. 1975, pp. 135-176.

I would like to thank David Weissman for reading an earlier version of the manuscript and offering a number of valuable suggestions, and Evan Fales for critical comments which helped me to clarify a number of points.

Although he might be surprised to hear it, this book would be rather different if it were not for the influence of Victor Lowe.

M. S.
Setauket, Long Island

Introduction

The purpose of this book is to investigate certain aspects of the ways in which concepts and theories in physical science may be related to one another. In particular, I shall be concerned with that relation which holds between two theories when one is said to be *reducible* to another. (For example, it is usually maintained that thermodynamics is reducible to statistical mechanics, and that Newtonian mechanics is reducible to Einstein's special relativistic mechanics.)

My primary goal is to move toward a general analysis of the concept of *theory reduction*. One of my main conclusions is that there are in fact two importantly distinct types of theory reduction: micro-reduction, which I shall refer to as *"concept replacement reduction"*— *theory* reduction via *concept* replacement, and *"direct theory replacement reduction."* The first is exemplified by the thermodynamics/statistical mechanics case; the second by the Newtonian mechanics/Einsteinian mechanics case. The first goes to a deeper level; the second stays at the same level.

The first type is discussed primarily in the first four chapters, where the development proceeds by first laying a foundation built with the aid of certain ideas which Bertrand Russell once employed in other areas: mathematics, logic, and metaphysics. My concept replacement analysis of this type of theory reduction in physical science is therefore introduced in a manner which shows its historical roots and its relation to similar reductive programs in areas other than physical science. I believe this manner of presentation to be of value because my analysis is importantly different from that found in current philosophic literature on theory reduction in science; thus the reader may find my alternative analysis more congenial if it is displayed in this historical and comparative manner. Moreover, it is of philosophic significance to notice that the concept of the reduction of one "theory" to another is a rather general one, with its occurrence in physical science being but one instance of its more general applicability.

The first two chapters contain the preliminary presentation of my analysis. Chapters 3 and 4 further elucidate my analysis of *concept replacement reduction* in the context of a comparison of it with what I call the "standard analysis" of reduction—an influential current view (or cluster of closely related views) which purports to be able to handle *all* examples of theory reduction in science. A number of related topics of importance are also discussed, for my analysis of reduction has significant implications for these related issues in the philosophy of science: the nature of *models*, the role of *laws* in science—particularly in theory reduction but also in explanation; the nature of *scientific explanation* and its relation to reduction; and the implications of a successful reduction for physical *ontology*.

Two general points emerge from this discussion: the centrality of *concept* replacement in theory reduction, and indeed the centrality of the concept of *replacement* itself—in terms of both *language* and *ontology*.

The second type of theory reduction which I distinguish—*direct theory replacement reduction*—is discussed in Chapter 5. This type of reduction will receive a briefer treatment than concept replacement reduction. There are a number of reasons for this. First, there is not as much need to motivate my analysis of this kind of reduction, for my analysis resembles the "standard analysis" to a greater degree than does my analysis of concept replacement reduction (even though it is in its application to cases of direct theory replacement reduction, oddly enough, that the standard analysis has received its most severe criticism in the current literature). Second, and in consequence of the above, there is no need to compare my analysis with the standard analysis at great length, since they are essentially the same in *structure*, differing mainly in the crucial notion of *replacement*. Finally, I have decided not to discuss at great length the specific *examples* which I give of such reductions, as I did in the case of concept replacement reductions, for such a detailed discussion (besides being unnecessary) would require a greater mathematical background on the part of the reader than I wish to presume. Moreover, the examples are probably already familiar, at least in outline, to the reader who gets this far in the book. Enough will be said, however, to establish that this is a quite distinct type of theory reduction. The relation between this type of reduction and the concept of a scientific *explanation* is also discussed.

In Chapters 6 and 7, I explore the idea of reduction as it seems to function with respect to items other than *theories*. Since my earlier analysis of concept replacement reduction trades heavily on what might appropriately be called *concept reduction*, it should come as no surprise that one can speak of the reduction of items other than theories. For example, one can speak directly of *entity reduction*, for it will emerge that there are in fact cases of reduction involving *no* "reducing theory" in the sense required by the standard analysis. This idea will be the subject of Chapter 6.

Of particular importance in the history of science is the idea of one entire *branch of science* being reduced to another (for example, chemistry to physics), which I discuss in Chapter 7. It is of interest that a proper analysis of this notion of *branch reduction*, in my view, bypasses the idea of *theory* reduction and is best approached by applying my notion of *concept* replacement (or concept reduction). If this proposal should turn out to be correct, I believe it would be an illustration of the power of my analysis of *theory* reduction on the basis of *concept replacement* in the first four chapters of the book.

The final chapter is devoted specifically to the assumption that one can always distinguish in a clear way between the terms and concepts of the reduced and reducing theory (or branch of science). This fundamental assumption seems to have gone unquestioned, not only by proponents of the standard analysis, but by its critics as well. I argue that this assumption is incorrect, and offer some suggestions regarding the implications of this realization, especially for the notion of branch reduction.

The issues which are considered central in the philosophy of science seem to change from generation to generation—as indeed they do in other areas of philosophy. Currently it would appear that the nature of *scientific change* is occupying a large number of philosophers (and historians) of science. This is evidenced, for example, by the great influence which Thomas Kuhn's *The Structure of Scientific Revolutions*[1] has had among philosophers of science since it appeared in 1962. It seems equally clear that one of the most important products of some types of scientific change is indeed *theory reduction*. For one of the most common types of major scientific advance involves the reduction of one theory (or central *concept*, or entire *branch* of science) to another. If this is so, then progress in understanding the concept of *reduction* in physical science would seem to be requisite to a full understanding of the nature of scientific *change*. It is my hope that this book will aid in advancing this goal.

It could also be argued that underlying the issues in philosophy which change from generation to generation are certain deeper ones which appear to be more durable. The search for conceptual simplicity and unity (and their ontological counterparts) in describing the world generates one such set of deeper issues. It is my belief that reductive programs in science are but specific examples of this more general and pervasive search which has perennially occupied philosophers. If this is so, then perhaps some of the thoughts presented in this study will also aid in clarifying some of these more general metaphysical problems.

Note

1. Thomas S. Kuhn, *The Structure of Scientific Revolutions*, Chicago, 1962; second enlarged edition, 1969. Indeed this book helped in large measure to *generate* the current interest in scientific change.

Concepts of Reduction
in Physical Science

Chapter 1

Russell's Maxim
and Reduction as Replacement

Russell's Maxim

In his paper "Logical Atomism"[1] Bertrand Russell stated a principle which he and Whitehead "found, by experience, to be applicable in mathematical logic,"[2] and suggested that it might also be of use in other areas. The principle is: "Wherever possible, substitute constructions out of known entities for inferences to unknown entities."[3] In his "Relation of Sense-Data to Physics," Russell stated the principle in this way: "Wherever possible, logical constructions are to be substituted for inferred entities."[4] He referred to this as "the supreme maxim in scientific philosophizing."[5]

In the opening chapters of this book, I want to clarify the meaning of this principle, which I shall refer to as "Russell's Maxim," and then follow Russell's suggestion to see what areas it can shed light upon. My primary goal is to use Russell's maxim to illuminate an important type of *intertheoretic reduction* within science (physics in particular). I shall be considering *microreduction*, which may also be labelled *concept replacement reduction*, for reasons which will become apparent. (In Chapter 5 I will consider another type of theory reduction which occurs in physical science—one which differs in important respects from the kind which I believe to be fruitfully analyzable on the basis of Russell's maxim.) I will begin with a clarification of the maxim, which will be accomplished with the aid of a survey of areas other than physics where it may be applied. Many of the things I will say are not new. But the approach and the particular combination of remarks which I shall make may, I hope, add to our understanding of the concept of intertheoretic reduction.

I shall begin by asking:

(1) What does 'logical construction' mean?

For definiteness, let us suppose that we have before us two theories (in a broad sense of the term—two bodies of statements about given domains, having some relatively clear deductive structure, but not necessarily fully axiomatized or formalized).[6] Call them 'A' and 'B', and let 'V_a' and 'V_b' refer to the vocabularies of non-logical constants appearing in each theory—the concepts characteristic of the theories.

Then, from "Relation of Sense-Data to Physics" it seems clear that what Russell meant by a logical construction of members of V_b out of members of V_a (say) is that each member of V_b is to be *explicitly defined* on the basis of V_a. ("We now define . . . , thus constructing it logically. . . .")[7]

In what follows I shall construe 'explicitly define' as 'replace', in a sense which shall become clear as I proceed. This would seem to conform with Russell's use of the term 'substitute' in the first statement of the principle above. It is this seemingly minor shift which will prove to be of importance later when I come to apply Russell's maxim to reduction in science.

We may also ask:

(2) What does 'wherever possible' mean?

Clearly, the principle may always be applied in a trivial manner; if there were no constraints, one could arbitrarily make up "definitions" of terms from V_b on the basis of terms from V_a. Therefore there must be some criteria which would allow one to distinguish successful from unsuccessful constructions in some interesting or important manner before Russell's maxim can be considered to be of any importance or interest. We will also have to discuss the nature of judgments of success or lack thereof, i.e., judgments as to whether or to what extent the criteria apply in any given case, as well as what we are to conclude from their application or (apparent) non-application in any given case.

For Russell, a successful construction must be such that the construct—an expression involving only members of V_a (as non-logical constants)—must have the same properties as the member of V_b for which it is being substituted; that is, *the construct must bear the same relations to other such constructs as did the original members of V_b to one another.* And this in turn seems to mean that any statement in theory B involving n terms of V_b which we hold to be true (were we to espouse B) must, when the n terms are *replaced* by their constructs out of members of V_a, be transformed into a statement in theory A which we hold to be true (were we to espouse A). (Notice that I have *not* said that the transformed statement must be *deducible* from the postulates of A. The importance of this will be seen when I come to apply Russell's maxim to reduction in science—where such a deduction is not always possible.)

A successful reduction is one which captures this aspect of the original usage (the usage in theory B) of the term being defined (replaced). The construction must supply what was once called a "real definition."[8] One could say that the gross structure of B must be reflected in A. The result of such a totally successful term-by-term construction and replacement will thus be the replacement of the entire theory B by (a portion of) the theory A. (It is thus apparent, and worth mentioning, that successful replacement of terms in this manner necessarily involves reference to the theories in which they are embedded.)

I wish to emphasize the *term-by-term* nature of the reduction. If intertheoretic reduction were not analyzed in this manner as a *term by-term* replacement, we would have the analysis of theory reduction given by Kemeny and Oppenheim in 1956.[9] They argued that the relation of reduction is best analyzed by comparing the theories in question directly with the "observations" they are intended to explain, rather than by attempting to draw any direct connection between the two theories themselves. They claimed, in essence, that B is reducible to A if A can explain any observations that B can explain. But in my view this weaker relation comes close to being *mere* theory replacement—not reduction. This will become apparent as we proceed.

A mathematical example may help to clarify some of these ideas. This will be an example of an application of Russell's maxim *within* mathematics. (The logicist reduction of mathematics to logic would be an example of an application of Russell's maxim *to* mathematics.)   My discussion of the example will be intentionally lacking in detail, because the procedural or methodological point involved is quite general.   Suppose that 'r' is a term which (purportedly) refers to some irrational number, such as the square root of two (an "inferred entity" in Russell's view). Russell's maxim bids us to (attempt to) substitute for 'r' some function of terms referring only to rational numbers (R_i, say) such that: if we let '$f(R_i)$' denote this function, and if $C(r)$ is some accepted statement in the theory of irrational numbers (our theory B in this example), then $C(f(R_i))$ will be an accepted statement in the theory of rational numbers (our theory A); and furthermore this will work for all C's—for any statement—in the theory of irrational numbers. (This example brings out the fact that theories A and B must have a great deal in common—in particular they must both be embedded in a common language. Thus, in the mathematical example, the function denoted by 'f' must be expressible in both theories, and the relations between terms implicit in the statement C must also be expressible in both theories. This will be brought out more clearly in the following pages.)

This last example will aid in answering a third question which we might ask in order to clarify Russell's maxim:

(3) What is meant by 'inferred entities'?

Or, what is the force of the distinction between 'known' and 'unknown' entities? In the mathematical example, what Russell had in mind is clear enough. He did not want to admit into his ontology two kinds of numbers—rational and irrational; rational numbers are enough (and even these are perhaps too much, and should eventually be eliminated in favor of classes—at least). But he did not want simply to jettison the theory of irrational numbers. (His reasons for this are not important for our present purposes.) If it could be shown, therefore, that irrationals can be construed as being at bottom only complex combinations of acceptable (or relatively more acceptable) rationals, we could have our cake and eat it too.[10] And we accomplish this by replacing all occurrences of r-expressions in the theory of irrational numbers by complex constructs involving only R-expressions in such a way that every accepted statement in r-theory becomes a (more complex) accepted statement in R-theory.

Having succeeded in this replacement program, we may keep the theory "of" irrational numbers and continue to use it, but without the assumption—the "inference"—that r-terms *refer* to anything in the universe other than (perhaps) the designata of the more complex R-expressions which have replaced them. The theory of irrational numbers may then be viewed or construed as a short-hand way of saying certain (rather complex) things about (classes of) rational numbers. Rather than assuming that the term ' $\sqrt{2}$ ' (e.g.) refers to the (irrational) limit of a certain sequence of rationals, we instead construe it as referring to a certain function of members of that sequence itself.[11] Irrational numbers (or irrational number expressions) have been *constructed* out of, and thereby *reduced* to, rational numbers (or rational number expressions).

Notice once again the *term-by-term* character of this procedure. We did not *merely* replace the theory of irrational numbers by some other theory; the relation between the two theories is much more intimate than that, for *each* "irrational number" was replaced by a function of rational numbers. It ought also to be noted here that it has not been demonstrated that "irrational numbers do not exist." The successful replacement program only shows that for certain theoretical purposes we need not assume that they do exist; or, paradoxically, one could argue that success shows that they *do* exist—although not as entities distinct from rationals. (I will have more to say about the existence of reduced entities in Chapter 4.)

There can be a variety of reasons which may motivate one to attempt to replace the vocabulary V_b of one theory B in a term-by-term manner with constructions out of the vocabulary V_a of another theory A in such a way as to preserve the original syntax of B as a portion of the syntax of A. The reasons will depend on a variety of factors such as subject matter, purposes, context, tradition, and also what might be called ontological ideology. Some of these matters will be clarified as we look at other areas in which one might attempt to apply Russell's maxim.

Russell's Maxim in Metaphysics

Russell's maxim, as he would apply it *to* physics (not *in* physics, which is part of my eventual goal), would have us replace matter by sense-data. I shall go beyond what Russell actually said in this regard and describe a possible two-step procedure of replacement.

Let 'T' refer to physical theory, 'V_t' its vocabulary (the set of "theoretical terms" used in physics); let 'P' refer to "physical object theory,"
by which I mean the set of ordinary statements we make about the
ordinary-sized objects we encounter every day (what Carnap once
called the "thing language"), and let 'V_p' refer to its vocabulary; finally,
let 'S' stand for sense-data theory, with V_s its vocabulary of terms referring to sense-data. (I am fully aware of the major difficulties involved in some of these notions, including the probable non-existence
of some of the purported "theories" I have just introduced. But the
reader will notice that in what follows this will not detract from the
methodological points I wish to make with their aid. For it is the
[ill-fated] *attempt* to apply Russell's maxim in this area which will
help to illuminate the idea of reduction.)

Now, *if* (for whatever reason) we wish to deny that theoretical terms
in physics designate items in the universe other than the ordinary molar
objects whose behavior is usually taken as the evidential basis for our
belief in their existence, and if we do not wish simply to throw out
physics as utterly meaningless or useless (who believes this?), we may
try to "substitute constructions out of known entities [the designata
of members of V_p] for inferences to unknown entities [suspicious
purported designata of members of V_t]." If we let an arrow standing
between two expressions mean '*shall be replaced by*', we may express
what Russell's maxim tells us to do in this way:

$$\text{For every } j, \ t_j \rightarrow f_j(p_1, p_2, ...) \tag{1}$$

where the $t_j(t_1, t_2,$ etc.) are members of V_t, and the p_i (p_1, p_2, etc.) are
members of V_p, and each f_j is some (complex) function of that which
appears within the parentheses. (What we have here, of course, is a way
of characterizing an early brand of what was once called *physicalism*—
one which proceeds in a term-by-term manner rather than, say, a state-
ment-by-statement manner.)

Similarly, *if* (for whatever reason) we wish to deny that physical
object terms ("thing-language" terms) designate items in the universe
distinct from sense-data, and if we do not simply wish to jettison our
ordinary talk about tables and chairs as being utterly meaningless or
useless, we may again try to "substitute constructions out of known
entities [the designata of members of V_s—sense-data] for inferences to
unknown entities [the alleged designata of members of V_p]." That is,

wherever some physical object term p_i occurs in sentences of physical object "theory" P, it is to be *replaced* by some appropriate function of members of V_s. Symbolically, as before:

$$\text{For every } i, \ p_i \to g_i(s_1, s_2, ...) \tag{2}$$

where, as before, the g_i are complex functions of what appears within the parentheses. This characterizes versions of *phenomenalism* which proceed in a term-by-term manner. (There are also versions of phenomenalism which proceed sentence-by-sentence,[12] and those which *do* simply jettison "physical object theory." The Kemeny-Oppenheim analysis of reduction would be more appropriate for this last form of phenomenalism.)

Expressions (1) and (2) may be combined to obtain:

$$\text{For every } j, \ t_j \to f_j(g_1(s_1, s_2, ...), g_2(s_1, s_2, ...), ...) \tag{3}$$

Using 'h' in an obvious way, expression (3) may be rewritten more briefly as

$$\text{For every } j, \ t_j \to h_j(s_1, s_2, ...) \tag{4}$$

The two-step replacement may thus be compressed, yielding theoretical terms as direct constructs out of sense-data. Russell's view when he wrote "Relation of Sense-Data to Physics" may therefore be analyzed and described as a combination of physicalism and phenomenalism (each of the term-by-term variety).

How is one to judge whether any specific attempt at one of the above constructions is successful? Let us consider the compressed version in (4) for simplicity.

Suppose that $C(t_1, t_2)$ is some true (or accepted) statement of physical theory, containing t_1 and t_2 as the only theoretical terms. The construction or replacement program schematized in (4) may be said to "look promising" if the sentence

$$C(h_1(s_1, s_2, ...), h_2(s_1, s_2, ...))$$

expresses a true (or accepted) statement in theory S about sense-data.[13] And the replacement is a total success if the same thing happens for *all* statements in T, each of which may contain any number of theoretical terms. For the truth-values of all statements of T would be preserved, but with no need to consider the members of V_t as designating anything (e.g., theoretical *entities*) other than complexes of sense-data which have replaced their purported designata. The general picture of a successful construction which emerges from this discussion of Russell's maxim may be put graphically as follows:

The theoretical *structure* or *syntax* of the original theory B (the theory being replaced) is preserved; but the nodes of this conceptual net, originally occupied by members of V_b, are now occupied by *structures*, *constructions*, composed of members of V_a. We therefore have before us (part of) theory A, and so we may continue to *use* the replaced theory B without ontological qualms, for its vocabulary can now be understood as shorthand for complex expressions in the replacing theory which, we suppose, does not raise the same ontological suspicions (although it may well raise different ones). And once again, the analysis of Kemeny and Oppenheim would have us pay no attention to the possibility of the reducing theory somehow *preserving* within it the syntactical structure of the reduced theory.

It is well-known that Russell's maxim can be used with great success within mathematics as well as in mathematical logic. Where else can it be successfully applied? Can the appropriate constructions be found for a reduction of physical object theory (expressed in Carnap's "thing-language") to sense-data theory? In my terminology, can a set of functions g_i be found for carrying out the replacement program expressed by (2)? It is now generally agreed that the answer to this question is *No*. This is just (one reason for) the failure of this type of phenomenalist program. (There are of course the other kinds of phenomenalism mentioned before.)

But I must hasten to add that there are ways in which one can "force" a success in this replacement program—or at least avoid or postpone failure. For example, one can simply discard those aspects or portions of "physical object theory" in which truth-values are not preserved under a set of replacement functions that do preserve the truth-values of other portions of physical object theory deemed more important for one reason or another. That is, one can treat preliminary successes as an indication that the program will in fact eventually succeed, and proceed to throw out anything that does not accord with these initial successes. Alternatively (and perhaps more respectably),

one can put aside for "further research" recalcitrant aspects of theory
B. These are familiar procedures for handling "difficulties" in many
areas, and not always to be despised. Each of these attitudes (and
variants of them) is always possible—indeed sometimes quite appropriate; and it seems to me that there are no general and precise criteria
for distinguishing perseverance from perversity in such situations. It is
an empirical fact today that philosophers who espouse phenomenalism
would be considered somewhat perverse in their espousal (although not
necessarily uninteresting). But this may well be a matter of current
style or taste.

Moreover, it would indeed be expecting too much to hope that *every*
statement of a reduced theory *B* should be transformed into a true
statement of the reducing theory *A*, because (at least in physical science, if not in every cognitive endeavor) the process of making and
testing constructs for the elements of V_h will usually unearth some
errors or misconceptions or infelicities of one sort or another in theory
B; it would not be reasonable to require error to be preserved in the
replacement program. The notion of a "totally successful" replacement
of which I spoke earlier must therefore be viewed as a kind of *ideal*, an
ideal that one does not usually expect to reach—especially in physical
science.

Can the appropriate constructions be found for a successful term-by-
term reduction of physical theory to physical object theory? I.e., can a
set of functions f_i be found which will effect the replacement expressed
in (1) above? Can talk about "theoretical entities" in physics be construed as shorthand for talk about tables and chairs and ammeters?
Again we know that the answer is *No*. The history of this unsuccessful
endeavor is well-documented.[14] We are all familiar with Carnap's
admission that theoretical terms cannot be explicitly defined on the
basis of "observation" (=physical object) terms.[15] In Russell's terminology, it has not been possible to find appropriate constructs to effect
the replacement. Even more "liberal" programs, involving (e.g.) Carnap's "reduction sentences," have failed. Very few philosophers now
attempt to "define"—in any reasonable sense of this abused term—theoretical terms on the basis of (physical object=) observation terms.
Indeed, the very attempt to distinguish the classes of terms involved in
the issue has come to be viewed as ill-founded. (See, e.g., my paper,
note 15.)

But here, too, failure need not be conceded; the procedures for avoiding or postponing failure which I described above may be—and have been—applied in this case. One can simply reject those statements of physical theory which are not transformed into acceptable statements of physical object theory, if the set of replacement functions seems to be otherwise satisfactory for those portions of physical theory considered for one reason or another to be of greater importance. (Gustav Bergmann once argued in essentially this manner.[16] He claimed that the idea of *explicit definition* could adequately handle dispositional concepts, even though certain statements from "ordinary language" were not thereby preserved. So much the worse for these statements! He therefore considered Carnap's retreat to "reduction sentences" to be unnecessary.)

It would be of interest to ask the converse question: Can physical object theory be reduced to physical theory? It could be argued that this is an apt description of one of the goals of *physics*, whereas the original reduction was once a goal of some philosophers. Russell's maxim may prove to be a useful tool of analysis in dealing with this question, but I will not pursue this issue here.

Notes

1. Russell, Bertrand, "Logical Atomism," reprinted in R. C. Marsh (ed.), *Logic and Knowledge*, Cambridge, 1956. (This paper was originally published in 1924.)

2. *Ibid.*, p. 326.

3. *Ibid.*, p. 326.

4. Russell, Bertrand, "Relation of Sense-Data to Physics," reprinted in Russell's *Mysticism and Logic*, London, 1917, p. 150. (This paper was originally published in 1914.)

5. *Loc. cit.*

6. As I proceed I shall introduce only that amount of detail, precision, and symbolism that the context requires. I also wish to avoid a situation in which possible criticisms of particular formalizations are thought of as thereby refuting the more general points which are my primary concern.

7. *Mysticism and Logic*, p. 150.

8. See, for example, p. 115 of Morris Weitz's contribution to P. A. Schilpp (ed.), *The Philosophy of Bertrand Russell*, New York, 1944.

9. Kemeny, John, and Paul Oppenheim, "On Reduction," *Philosophical Studies*, Vol. 7 (1956), pp. 6-19. Reprinted in Brody, B. (ed.), *Readings in the Philosophy of Science*, Englewood Cliffs, 1970, pp. 307-318.

10. In Chapter 4, I will discuss in more detail the situation in which we are able to do without a theory in principle, but keep it in practice.

11. I have been implicitly using the theory of the Dedekind Cut. For more detail of a philosophically informed character, see Russell's *Introduction to Mathematical Philosophy*, London, 1919, Chapter VII; and David Hawkins, *The Language of Nature*, San Francisco, 1964, p. 70 ff.

12. See, for example, Hans Reichenbach, *Experience and Prediction*, Chicago, 1938.

13. Note once again that it is here implicitly assumed that the structure expressed by C can occur both in physical theory and in sense-data theory. I am thus assuming that the sentential function C is not a term from the *specific* vocabulary of T; indeed, if it were, it too would have to be replaced by terms *only* from S. I am therefore short-circuiting what would really amount to a several-layered analysis and assuming that C involves only a vocabulary common to T and S. The reader can supply a sketch of what is involved in these "layers"; I have omitted this in order to avoid a forest of symbols which would only obscure what I am trying to make clear.

14. See, for example, the opening chapters of Arthur Pap's *An Introduction to the Philosophy of Science*, New York, 1962; or Carl Hempel, *Fundamentals of Concept Formation in Empirical Science*, Chicago, 1952.

15. See R. Carnap's 1936-37 paper "Testability and Meaning," reprinted in H. Feigl and M. Brodbeck (eds.), *Readings in the Philosophy of Science*, New York, 1953. (Carnap's identification of observation terms with physical object terms from the "thing-language" is discussed in my paper "Theory and Observation," *British Journal for the Philosophy of Science*, May, 1966.)

16. See G. Bergmann's "The Concept of Cognitive Significance," *Proceedings of the American Academy of Arts and Sciences*, 80 (July, 1951), 78-86; reprinted in G. Bergmann, *The Metaphysics of Logical Positivism*, New York, 1954.

Chapter 2

Russell's Maxim
and Reduction in Physics

So far then, it appears that outside of mathematics and mathematical logic the fruits of Russell's maxim are meager. (And, after all, one can find very few references in the current literature to Russell's work in this area.) I would like to show, however, that Russell's maxim plays a powerful role in an area we have not yet examined. I shall argue that it is a very general, implicitly used tool *within* physics. That is, a certain pervasive type of intertheoretic reduction *in* physical science—*microreduction*—can be illuminated by an examination from the point of view of Russell's maxim; physical science contains what I have called *concept replacement reductions*. In the present chapter, I shall attempt to show this in some detail for one concrete example. In the following two chapters, I will discuss in more general terms the advantages and implications of thus viewing this type of intertheoretic reduction.

I shall be considering the relation between Maxwell's electrodynamics and classical Newtonian dynamics, as conceived during the latter part of the nineteenth century. I will first sketch, in a form suitable for my purposes, certain aspects of the structure and function of these two theories. A more detailed description may be found in my book *Methodological Foundations of Relativistic Mechanics* (Notre Dame, 1972).

Mechanics and Electrodynamics

Newtonian dynamics deals with the kinds of motions a system of massive bodies will undergo when under the influence of different kinds of forces. Newton's second law of motion states that whenever a force **F** acts upon a body of mass m the result will be that the body will accelerate in accordance with the equation

$$\mathbf{F} = m\mathbf{a}$$

In any particular situation, the force acting upon a given body will depend on various parameters describing the body and the environment in which the body finds itself—its mass, electric charge, distance from other bodies, velocity of the body with respect to others in its vicinity, various coefficients, etc. One can write down particular *force laws* for particular kinds of situations which describe exactly how the force is determined in accordance with these parameters. When one of these particular force laws is combined with the general law **F** = $m\mathbf{a}$ and the resulting equation is integrated, we can calculate (and thus explain) the motion of the body due to the force operative in the particular situation. Here is a partial list of some of these force laws (most of which I have put in scalar form):

$f = G\,\dfrac{mM}{r^2}$	Newton's law of universal gravitation
$f = mg$	Galileo's "law of gravity"
$f = -kr$	Hooke's law (the "law of springs")
$f = -kr - c\,\lvert v \rvert$	Hooke's law with friction
$f = \dfrac{e_1 e_2}{r^2}$	Coulomb's law of electrostatic attraction
$\mathbf{f} = e\left\{ \mathbf{E} + \dfrac{1}{c}\,(\mathbf{v} \times \mathbf{H}) \right\}$	Lorentz force law from electrodynamics

Laws such as these, together with the general force law **F** = $m\mathbf{a}$, are used to explain the motions of bodies in situations covered by these laws.

Each of the above force laws characterizes a *branch* of classical mechanics. The first one yields classical gravitational theory, which explains the motions of massive bodies due to their mutual attraction as a function of their masses and relative distances. The third—or certain generalizations of it to three dimensions—yields the mechanics of elastic bodies ("Hookian solids").

The last law of this partial list (together with Maxwell's equations, which I shall presently introduce) yields that branch of mechanics known as *electrodynamics*. It may surprise some readers that I have referred to electrodynamics as a "branch" of classical Newtonian dynamics, since the usual story is that the two are quite distinct theories. But I wish to emphasize that the Lorentz force law is methodologically on a par with the other force laws listed. That is, it too expresses a function of various aspects of the nature and environment of a body which, when plugged into the left side of $\mathbf{F} = m\mathbf{a}$ and integrated, will yield the motion of the body in the environment in question. As such, the use of the Lorentz force law conforms to a standard pattern of explanation found in classical mechanics. In this regard, it differs from the other force laws only in its relative complexity, including its functional form and the occurrence of a larger number of parameters of the environment. Aside from this, it is fundamentally on a par with (say) Hooke's law with friction, or any other law on the list.

One further difference that might tempt one to resist interpreting electrodynamics as a branch of mechanics is the occurrence of certain further equations involving the same parameters which enter into the Lorentz force law. I have in mind Maxwell's equations, one form of which is:

$$\nabla \cdot \mathbf{E} = \rho \, , \qquad\qquad \nabla \cdot \mathbf{H} = 0 \, ,$$

$$\nabla \times \mathbf{E} = -\frac{1}{c}\frac{\partial \mathbf{H}}{\partial t} \, , \qquad \nabla \times \mathbf{H} = \frac{1}{c}\left(\frac{\partial \mathbf{E}}{\partial t} + \rho \mathbf{v}(\rho)\right) .$$

There is no similar set of equations accompanying Hooke's law. But this difference too is only one of relative complexity. For Maxwell's equations are best interpreted methodologically as *equations of constraint* (as this term is used in analytical mechanics)—equations which relate some of the parameters which enter into the force law in question, in this case the Lorentz force law. If there were similar equations appended to Hooke's law with friction, relating k, c, and v, this would not suffice for considering elasticity theory to be a theory distinct from classical mechanics—it would still be a *branch* of the latter; I therefore see no reason to draw a different conclusion from the fact that such equations do occur in the case of electrodynamics. As a matter of fact, a full statement of the mechanics of elastic bodies—a generalization of Hooke's law—*does* contain equations of constraint and is fully as rich as electrodynamics in its mathematical complexity; and it is clearly a (venerable) branch of classical mechanics.

Reduction of Electrodynamics

This has been a digression, but an important one, for I began by listing various force laws, including that of Lorentz, and I have considered each such law to give rise to a branch of classical mechanics. It might appear then that classical mechanics will have many branches, given by the number of force laws we can generate and add to the list—too many branches, in fact, for the entire explanatory function of the theory would be trivialized if no limits were placed on the number and kind of such force laws to be admitted to the list. This is because, for any given situation, one can always generate in a trivial manner a *mathematically* satisfactory *formula* which, together with $F = ma$, will predict the motion of a given body, by reading back to the formula, as it were, from the observed motion of the body. But not every such "engineering formula" is considered to be a *force law* that *explains* the motion of a body. (Further details on this and similar points may be found in my book, previously mentioned. On this specific point, see pp. 36 ff.)

Therefore classical mechanics becomes an interesting, unifying explanatory theory only when certain limitations are placed upon the set of formulas and their associated parameters considered to be *force laws*, which, then, serve to *explain* the motions of bodies. In particular—an important point for our purposes—one attempts to *reduce* "engineering formulas" to force laws, that is, to reduce mathematically satisfactory formulas to others which are considered somehow more basic, simpler, more intelligible, or more "purely mechanical"; generally, they are more in accord with our beliefs (developed from past experience) regarding the basic kinds of forces actually operative in the universe and the basic constituents of the universe designated by the parameters entering into these laws. (Again, see pp. 36 ff. of my book on relativistic mechanics for details.)

It is previous success with a limited number of such simple force laws involving interpreted[1] parameters that have become familiar (and thus taken to refer to "known entities") which determines that in future problems we attempt to use only (combinations of) these functions and parameters as *force laws* to *explain* the motions of bodies. We attempt to *treat (apparently) new ("unknown") environments as really being the old ("known") kinds disguised in complexity.* (The reader may foresee the burden of the next section by comparing this last sentence with the statement of Russell's maxim early in the previous chapter.)

I said above that the Lorentz force law (and the associated Maxwell equations) give rise to electrodynamics as a *branch* of mechanics. I also said that the complexity of these equations should not blind one to this methodological fact. But this same complexity led some physicists to suspect that the Lorentz force law is one of those *un*simple, *un*basic, *un*intelligible laws which should be struck from the list of basic *force* laws along with its distinctive parameters—that it should be viewed as an "engineering formula."[2]   It was believed that one should attempt to reduce the laws of electrodynamics to others on the list of force laws (together with any associated equations of constraint). And now we can apply Russell's maxim directly to describe this particular reductive program *within physics*. I will show in detail how Russell's maxim can illuminate this example.

Reduction and Russell's Maxim; Reduction and Models

Let M be the set of mechanical parameters that enter into the force laws on the list other than the Lorentz force law. Let ED be the set of parameters distinctive of electrodynamics—Maxwell's equations and the Lorentz force law, i.e., the set $(\mathbf{E}, \mathbf{H}, \rho, e, c)$.[3]   Russell's maxim tells us to attempt to *replace* each member of ED with a construct—a function of—members of M in the equations of electrodynamics. The replacement program will be a complete success if all previously accepted statements of electrodynamics (statements containing terms from the set ED) are, upon replacement of terms, transformed into accepted statements from the remainder of mechanics—statements containing only terms from the set M.

Maxwell's equations are of course "previously accepted statements" from electrodynamics. Let us consider, for example,

$$\nabla \times \mathbf{E} = -\frac{1}{c}\frac{\partial \mathbf{H}}{\partial t} . \tag{1}$$

Russell's maxim directs us to attempt to construct functions $\mathbf{g}_1, \mathbf{g}_2$, and $\mathbf{g}_3$ of members of M such that, upon carrying out the *replacements* expressed by

$$\mathbf{E} \to \mathbf{g}_1 \,(\text{members of } M),$$
$$\mathbf{H} \to \mathbf{g}_2 \,(\text{members of } M),$$
$$c \to g_3 \,(\text{members of } M),$$

we will have, in place of equation (1) from electrodynamics, the equation

$$\nabla \times \mathbf{g}_1 = -\frac{1}{g_3}\frac{\partial \mathbf{g}_2}{\partial t} \tag{2}$$

where this last equation is an accepted statement *in (the remainder of) classical mechanics*—the reducing theory in this example. By 'accepted', it is not meant to be required that this equation already exist "on the books" of classical mechanics. The equation may never have been seen before. It would be sufficient if this equation—this statement about how various mechanical parameters are related—could be shown to follow from other already accepted statements in mechanics (from other force laws on the list plus their equations of constraint, for example). But it is important to notice that even this is not necessary, as I will argue in later chapters.

Consider also the Lorentz "force law" itself:

$$\mathbf{f} = e\left\{\mathbf{E} + \frac{1}{c}\,(\mathbf{v} \times \mathbf{H})\right\} \tag{3}$$

We must, in accord with Russell's maxim (which expresses the reductive program) find an additional function g_4 of members of the set M such that the replacement of the concept of electric charge e by g_4, together with the original three replacements above, will transform the Lorentz equation (3) into:

$$\mathbf{f} - g_4\left\{\mathbf{g}_1 + \frac{1}{\mathbf{g}_3}\,(\mathbf{v} \times \mathbf{g}_2)\right\} \tag{4}$$

And this must be an accepted equation within the remainder of classical mechanics.[4] It would thereby be shown that alleged electromagnetic forces given in equation (3) are not fundamental, but are simply complexes of other types of mechanical forces given in equation (4). The Lorentz formula would no longer be a fundamental *force law*.

If this program could be carried out—if every statement in electrodynamics[5] could thus be replaced by statements from the remainder of mechanics when the electrodynamic parameters are replaced by functions of other mechanical parameters, it would thereby be shown that the specifically electrodynamic parameters of the set *ED need not be construed as designating any entities in the universe distinct from the designata of the constructs which have replaced them.* They may instead be viewed as shorthand for those more complex expressions. 'E', for example, rather than being interpreted as designating an "electric field intensity," could be considered instead as shorthand for a complex expression referring to (e.g.) the mechanical strain in some pervasive elastic medium (if that is what the function g_1 turns out to express). Alternatively it could be said that so-called electric fields have been shown to be nothing but such mechanical strains. And '*e*', rather than being considered as referring to a distinct property of bodies known as their "electric charge," could be looked upon as shorthand for the complex expression referred to by 'g_4' designating (e.g.) a mechanical vortex motion in this medium. Alleged electric charges are, after all, nothing other than vortices in the ether.

Notice how this kind of *replacement* is quite different from *rejection*: this is not a "mere" replacement of electrodynamics with another theory, but a replacement in which it is required that certain features of the replaced theory be preserved within the replacing theory. We do not simply use the remainder of mechanics in place of electrodynamics for the original domain of electrodynamics, but we use mechanical laws to *mimic* the electrodynamic laws by preserving their structure. The Kemeny-Oppenheim analysis of reduction, about which I spoke in the previous chapter, does not capture this important feature of the kind of intertheoretic reduction under consideration.

I am tempted to express this difference between *mere* replacement (of the Kemeny-Oppenheim variety) and *reduction* (as I have analyzed it) by saying that a successful *reduction* in this case would show how mechanical concepts and laws can be used to *model* electrodynamic processes. But I am afraid that this might be understood in the sense of a "mere" model or analogue, whereas my intent is to treat a successful reduction in a *realistic* manner. A successful reduction usually leads to dropping the "model" terminology. Phenomena previously accounted for with the aid of the reduced theory are, subsequent to a successful reduction, simply considered as being *constituted* by, and explained on the basis of, the entities and laws of the reducing theory.[6]

It might be useful at this point to take a moment to forestall a possible misconception. One must not be misled into believing that all that has been accomplished here (assuming a successful reduction) is to show that two otherwise totally distinct theories (mechanics and electrodynamics) happen to have certain *purely formal* characteristics in common. As I have argued elsewhere,[7] various theory-pairs in physics may be shown to be related in this weaker way, e.g., selected portions of acoustical theory and electric circuit theory. But of course we don't therefore claim a reduction in these cases. We don't claim, e.g., that sound vibrations are *composed of* electrical entities, or that the electrical resistance in an automotive spark plug wire (e.g.) is really at bottom a complex combination of unobservable sound waves in the wire.

But the replacement procedure described above in the mechanics-electrodynamics case does more than show how two otherwise distinct theories happen coincidentally (and thus inconsequentially) to share certain *purely* formal or structural characteristics. For these theories also have a domain of physical phenomena in common. There is a large measure of overlap in the domains of observable phenomena they are invoked to explain. In this case of reduction, the mechanical laws are used to explain the same phenomena which the electrodynamic laws explained. There are thus what I once called *substantive similarities* as well as formal ones.

For example, some of the lower-level laws which Maxwell's equations and the Lorentz force law are invoked to explain deal with damped vibratory motions in an elastic medium—travelling sinusoidal waves that decay with time (the "electric waves" in the "ether" which Hertz correctly predicted). And the *mechanical* equations which are generated from the electrodynamic laws in the manner I have described in my analysis also have, as lower-level deductive consequences, not merely *formally* similar laws but also laws which describe damped vibratory motion in a (complex) elastic medium. It is this shared lower-level domain—the shared data-base, together with the shared formal structure—which allows one validly to argue, in the above manner, that the relevant electromagnetic entities are actually *composed of* the complexes of related mechanical entities. Success in the reduction would thus mean that mechanical phenomena no longer *merely* model electrodynamic phenomena; they actually *constitute* the latter.

It is tempting to agree with Berent Enc's

> . . . conjecture that any theory that can be shown to contain a substantive model is an example of intertheoretic [micro]reduction

and, conversely, that in any case of intertheoretic microreduction
a substantive model can be found.[8]

The problem with this is that if one looks at the notion of a "substantive model," the "conjecture" (apparently meant to require further
research to substantiate) becomes true, but tautologically so. For in
my paper on models, to which he refers, items which he calls "substantive models" (Enc's term) are, *if totally successful*, no longer (merely)
"models"; they simply become correct descriptions of underlying entities, structures, and processes. His "conjecture" boils down to the
fact that successful theories containing descriptions of underlying
entities (Enc's "substantive models") are microreductions; conversely,
microreductions contain descriptions of underlying entities ("substantive models")—something a good dictionary might have told us.

I offer the following as my own conjecture regarding the relation
between *microreduction* and *model building*: The term 'model' is often
used in those cases in which an attempted reduction, because of the
nature of evidence, has not *yet* been accepted or rejected. ("It is only
a model *as yet*.") If further empirical evidence is negative and leads to
the rejection of the attempted reduction, we obtain another common
use of the term 'model'. ("It was, after all, *only* a model.") If further
empirical evidence leads instead to acceptance of the attempted reduction, we usually *no longer* refer to it as a model—it is too good, as it
were. ("It is no longer *just* a model.") The term 'model' then, is used
to describe preliminary or early attempts at microreduction, or failed
reductions (which may in spite of ultimate failure be quite useful for
certain purposes). And all of this *is* a "conjecture"; its truth would
have to be shown on the basis of a detailed examination of scientists'
actual usage of the terminology involved.[9]

What I have been describing with the aid of Russell's maxim are, of
course, the nineteenth-century attempts to construct what were called
"mechanical *models* of the ether." I have been trying to show the manner in which Russell's maxim can clarify the logic of these *attempted*
but *unsuccessful* reductions—these *models*. (In the next chapter I will
compare my way of viewing this kind of intertheoretic reduction with
the kind of analysis usually given in recent discussions of reduction in
the philosophic literature.)

In this *concept replacement reduction* (albeit one that did not succeed), the "known entities" of Russell are the *recognized* forces and

parameters of (pre-electrodynamics) mechanics in *new combinations*. The "unknown entities" are the supposed ("inferred") referents of the terms 'E', 'H', etc. taken as being distinct and independent aspects of physical reality. The "constructions" are in this case mathematical functions of the *old* mechanical parameters which behave in a way which satisfies the new electrodynamic equations.

The success of such a program in this case could be described in the graphic terms used in Chapter 1 as follows:

The mathematical structure of electrodynamics would be preserved; but the nodes of this mathematical-conceptual net, originally occupied by electrodynamic parameters, would now be occupied by *structures— constructions*—composed of mechanical parameters from the remainder of mechanics, the set *M*. We would therefore have before us (part of) pre-electrodynamics mechanics, and we could therefore continue to *use* electrodynamics without ontological qualms, for its distinctive vocabulary could now be understood as shorthand for complex expressions in the remainder of mechanics which, we suppose, do not raise the same ontological suspicions (although of course other such suspicions may arise).

To remove any doubts about the literalness of the *attempts to reduce* electrodynamic parameters to (already known) mechanical parameters— attempts to treat Maxwell's equations (e.g.) as ". . . mechanical consequences of concealed structure in that medium [the ether] . . . ,"[10] I would like to put on record this description of one such attempt:

Another device . . . was described in 1885 by FitzGerald; this was constituted of a number of wheels, free to rotate on axes fixed perpendicularly in a plane board; the axes were fixed at the intersections of two systems of perpendicular lines; and each wheel was geared to each of its four neighbors by an indiarubber band. Thus all the wheels could rotate without any straining of the system, provided they all had the same angular velocity; but if some of the wheels were revolving faster than others, the indiarubber bands would become strained. It is evident that the wheels in this model play the same part as the vortices in Maxwell's model of 1861-2: their rotation is the analogue of magnetic force; and a region in which the masses of the wheels are large corresponds to a region of high magnetic permeability. The indiarubber bands of FitzGerald's model correspond to the medium in which Maxwell's vortices were

embedded; and a strain on the bands represents dielectric polariza-
tion, the line joining the tight and slack sides of any band being the
direction of displacement. A body whose specific inductive capacity
is large would be represented by a region in which the elasticity of
the bands is feeble. Lastly, conduction may be represented by a
slipping of the bands on the wheels. . . .[11]

It should also be noticed how the term-by-term nature of the replacement is clearly illustrated in the latter half of this passage. Each concept
from electrodynamics is replaced by a complex construct of mechanical
concepts.

It is well-known that the reductive program in the case of electrodynamics failed. This partly accounts for the way in which Whittaker (and
the physicists of the time) speak only of "models" or "analogues" rather
than in a more realistic manner. The replacement did not work, and it
failed for empirical reasons. In general, such replacement attempts
were *only* models, because they could only handle *some* situations. In
the terminology of my analysis, each attempted set of replacement
functions (and many possible such sets were offered) transformed at
least some important electrodynamic laws into statements couched in
the language of mechanics which were shown empirically to be false.
Various hypotheses that the electromagnetic ether was at bottom this
or that type of mechanical system turned out to be falsified. But each
such hypothesis, represented by a set of replacement functions, *could*
correctly handle some other subset of the relevant observed phenomena. Hence each set of replacement functions represented *only a model*.
(I have discussed some of these matters in more detail in my paper referred to in note 6; see especially section 2.)

If the reductive program had succeeded, it would have been shown
that electrodynamics is doubly a branch of mechanics. For not only
would it be a branch of mechanics in the methodological sense which I
discussed earlier, but also in the further sense that the electrodynamic
equations would be at bottom part of (perhaps even deducible from)
another branch or branches of mechanics.

I merely note now that current discussions of reduction in physics
emphasize deducibility at the same they relegate the term-by-term constructions that I have been considering to "bridge law" status. My

analysis in terms of Russell's maxim emphasizes these constructions and will eventually place the deducibility of laws in the background as a test—though not a necessary test—of the success of these constructions. Before I go on to expand on this and related points in the next chapter, I shall conclude this chapter by comparing the failure of the reductive program for the physical case of electrodynamics with its failure in the metaphysical cases of phenomenalism and physicalism.

The most obvious difference between the failure of phenomenalism (or that of physicalism) and the failure to reduce electrodynamics to the remainder of mechanics is, of course, that the latter failure is empirical in nature. One could not have known in advance of empirical investigation that the reduction would not succeed; whereas it is not an empirical matter that the phenomenalist and physicalist reductions failed.

But this is an oversimplification; for to leave the matter thus is to cover up some important similarities between the two cases. In the physical situation as well as the metaphysical, there was a time during which investigators interpreted partial successes as indicative of eventual complete success; and refutations (as they would *now* be viewed) of particular attempts were looked upon as "difficulties" to be overcome by further research.

The point is that the testing of any proffered set of replacement functions in order to see if such a set satisfies the relevant criteria of success is not at all an easy matter. (It may be remarked quite generally that criteria may be clear, yet judgments as to whether they apply in a given case may be difficult to make; and decisions as to what conclusions to draw from their apparent non-application may be more difficult still.) For under various attempted sets of replacement functions, the electrodynamic equations were transformed into quite complex mechanical equations meant to apply to a submicroscopic domain of mechanical elements; also, the hypotheses about the detailed structure of the ether which these equations encapsulated could by no means be directly tested. Often only deductions far down the computational line, involving various mathematical approximations, were required before one had a statement that could be "directly confronted" with experience—and the "experience" in question often involved small effects detectable only with crude instrumentation, though the theory of the instruments required might well involve some quite esoteric branches of physics, applicable only with some difficulty.

It should come as no surprise, then, that various sets of attempted replacement functions could each meet with partial success in various subdomains in overlapping ways, so that it would be very difficult indeed to judge in a non-arbitrary way which set was more promising than another, or indeed whether (or when) the entire enterprise should be continued or abandoned.[12]

Therefore, here too, as well as in the metaphysical situations discussed in the previous chapter, it is difficult to distinguish perseverance from perversity when it comes to judging repeated attempts along one line or another, or when it comes to deciding whether to continue the reductive program in general. There does come a time, however, when a growing number of investigators simply get tired of the effort; those who persevere then are considered by the rest perverse. At this point, one finds "philosophical considerations" coming to the fore to "explain" why the entire endeavor was mistaken *a priori*, arguments based for instance on the nature of explanation in science[13] or the nature of theoretic reduction or arguments about ontology quite similar to those one finds in the metaphysical disputes about phenomenalism or physicalism.

There is another respect in which these paired failures are similar—a feature I shall only mention briefly. It is in a sense the converse of the considerations just offered; for now, it is the physical case itself which is said to have an important *a priori* characteristic. The point is that classical electrodynamics is a *relativistic theory* whereas the remainder of mechanics is not. That is, the equations of electrodynamics are invariant under a *Lorentz* transformation between inertial frames of reference whereas those of the remainder of mechanics are invariant under a *Galilean* transformation. (See my book, *op. cit.*, p. 109.) Now since (at most) only one of these modes of translating descriptions from one inertial frame to another can be correct, we have what may be deemed a non-empirical reason for concluding that the reduction attempt *must* fail—a reason which could not be clearly understood before the advent and acceptance of Einstein's special theory of relativity in the early twentieth century. This is somewhat similar to showing that phenomenalism *must* fail because of some sort of "category mistake" (say) involved in the very attempt to reduce physical objects to sense-data.

Notes

1. See my book (cited in the text), p. 38.

2. There are other physical reasons which motivate this attitude toward the Lorentz force law, but to give them here would entail too large a digression into the details of late nineteenth century physics. My purpose throughout this chapter is to delineate a way of analyzing the structure of a kind of reduction, whatever its physical motivation may be.

3. There are other ways of formulating electrodynamics in which this exact set does not occur. But those who are informed enough to realize this are probably astute enough as well to see that it does not matter for present purposes.

4. We now have some concrete examples of what I referred to in Chapter 1 as a "common language" for two theories where one is being reduced to the other. Comparing equations (1) and (2), and equations (3) and (4), it is clear that certain (mathematical) expressions such as '∂/∂' and 'x' ("curl") occur in both mechanics and electrodynamics. It is the language of mathematics which provides this common language or matrix for applications of Russell's maxim within physics.

5. If the theories are axiomatized it would be sufficient to do this for each of the axioms of electrodynamics—but not necessary, as we shall see later; indeed, it is not feasible to do so, though physicists try.

6. See my paper "Models and Theories," *British Journal for the Philosophy of Science*, Aug. 1965, pp. 121-142; reprinted in B. Brody (ed.), *Readings in the Philosophy of Science*, Englewood Cliffs, N.J., 1970, pp. 268-293. Section 3 of the paper is especially relevant to the point at issue.

7. *Ibid.*

8. Berent Enc, "Identity Statements and Microreduction," *Journal of Philosophy*, Vol. 72, June 10, 1976, p. 305. He uses the term 'substantive model' to refer to the work of R. Harré, *Theories and Things* (New York, 1961); M. Hesse, *Models and Analogies in Science* (Notre Dame, 1966); and my own (*loc. cit.*).

9. The beginnings of such an investigation may be found in my paper on models, and in P. Achinstein's papers, "Theoretical Models," *British Journal for the Philosophy of Science*, 16 (1965), pp. 102-120, and "Models, Analogies, and Theories," *Philosophy of Science*, 31 (1964), 328-350.

10. This is J. Larmor, writing in 1900, as quoted by E. T. Whittaker, *A History of Theories of Aether and Electricity*, New York, 1960, Vol. 1, p. 303. Larmor was *against* attempts of this kind. See note 13.

11. Whittaker, *op. cit.*, p. 292.

12. It should be clear that current formal theories of confirmation would be utterly useless in such a situation (but not for this reason unimportant for understanding the abstract nature of confirmation—or at least certain features thereof).

13. Consider, e.g., Larmor again (in Whittaker, *loc. cit.*): "We should not be tempted towards explaining the simple group of relations which have been found to define the activity of the aether by treating them as mechanical consequences of concealed structure in that medium; we should rather rest satisfied with having attained to their exact dynamical correlation, just as geometry explores or correlates, without explaining, the descriptive and metric properties of space."

Chapter 3

Concept Replacement vs.
the Standard Analysis of Reduction—Part I

I have now completed the preliminary presentation of my analysis of one important type of intertheoretic reduction—microreduction. The main idea has been that in these reductions the primary focus is on an *item-by-item replacement* of each of the concepts (or terms, or entities in fact) of the reduced theory by *complexes* of such items from the reducing theory. As an example of reductions of this type, I have used an unsuccessful but nevertheless important such attempt from the history of science. Indeed the lack of success has helped to shed some light upon certain features of the reductive enterprise, e.g., its relation to the concept of a *model*. (In Chapter 4, I will discuss a successful concept replacement reduction.)

In this chapter and the next, I wish to elucidate further my analysis by comparing it with another approach which, in recent years, has become a kind of "standard analysis" of reductions of all types.

I shall begin with an outline of this standard analysis. Then I shall approach the comparison via a criticism of some points of the standard analysis. It may be found, prominently, in the writings of Carl Hempel and Ernest Nagel.[1] It may be ultimately incorrect; in fact it has been recently attacked on a variety of grounds. But at the very least it provides a useful background for presenting and evaluating alternatives. In brief outline, it is this:

The Standard Analysis

The reduction of one theory B to another A requires two things: (1) the reduction of the *laws* of B to the laws of A, and (2) the reduction of the *terms* or *concepts* of B to those of A.[2] (It should be readily apparent that this analysis, like mine, requires more than Kemeny and Oppenheim required in their analysis of reduction. For here, too, the theories in question are required to be related to *each other* in specific ways rather than only to their data bases.) That is, in order correctly to claim that a theory B has been reduced to a theory A, two conditions must be fulfilled. The first one is:

(i) The postulates or basic laws of B must be logically *deducible* from the postulates of A.

Such a deduction requires the aid of a certain set L of "connecting principles" or "bridge laws" each of which contains terms from both vocabularies V_a and V_b. Thus, strictly speaking, we do not deduce the laws of B from the laws of A alone. It *should* therefore be said that B must be deducible from the *conjunction $A \cdot L$*, hence is *reducible to this conjunction*.

The second condition is one involving these bridge laws and the *terms* of the two theories. It is:

(ii) Each term of B must occur in a bridge law of *biconditional* form with terms of A.[3]

That is, each term in V_b must be correlated with a set of terms in V_a via a bridge law of the following form (in which I let 'T_b' stand for any term from V_b):

$$---T_b--- \text{ if and only if } \ldots,$$

where '$---T_b---$' is a statement in which T_b is the only term from V_b, and where '. . .' is a statement in which the only descriptive terms are from V_a. This, then, is the "standard analysis" of reduction, which is claimed to apply to *all* cases of theory reduction in science.

As Hempel has pointed out, it is essential that the bridge laws be of *biconditional* form, for reduction of *laws* in the sense conveyed in condition (i) is not a sufficient condition for the reduction of the relevant *theories*. Hempel's precise reason for this claim is of some importance. It is that

. . . short of a [corresponding reduction of concepts], the reduction of [B to A] would amount to the establishment of a set of connecting principles which . . . would constitute additional [laws of B] and would thus simply expand the theory B, but would not make its conceptual apparatus dispensable.

Full reduction of concepts in this strict sense would require, for every term of B, a connective law of biconditional form, specifying a necessary and sufficient condition for its applicability in terms of concepts of $[A]$ alone. Such a law could then be used to 'define' the $[B]$ term and thus, theoretically, to avoid it.[4]

In terms of the standard analysis, I agree fully with Hempel's claim that the reduction of the *theory* B to the theory A would require bridge laws of biconditional form. But his specific reasons for this claim lead to some unacceptable results. It will be instructive to see why this is so, for it leads directly to the advantages of my alternative analysis.

'L' stands for the conjunction of the bridge laws which are required in order to achieve a deduction (and hence reduction) of the laws of B from those of A. That is, 'B' can be deduced from the conjunction '$A \cdot L$'. This deduction constitutes the reduction of *all* of the laws of B to those of A—it is simply the fulfillment of condition (i). Now suppose that some of the bridge laws constitutive of the set L are *not* of biconditional form.[5] According to Hempel, these non-biconditional bridge laws must be considered as being additional laws of B (they would "simply expand the theory B"). But if this is so—if L consists at least partly of statements from theory B, then these statements too would have to be reduced to the theory A, since we are assuming that all of the laws of B have been reduced to those of A (condition [i]). But this in turn means that there must be further bridge laws—call them L'—which in conjunction with the laws of A will allow for the deduction of the non-biconditional constituents of the original set of bridge laws L. If we now assume that some of the statements constitutive of L' are themselves non-biconditional, the regress involved becomes apparent. This regress will stop only if at some stage we reach a set of bridge laws L^i all of which are biconditionals. It therefore follows that *if* reduction of *laws* has been accomplished, then there must be some bridge laws (the set L^i) of biconditional form. I do not think that Hempel would wish to accept this conclusion, which says in effect that satisfaction of condition (i) *entails* partial satisfaction of condition (ii). For Hempel himself gives an example of law reduction involving only non-biconditional bridge laws.[6]

Moreover it is at least odd to claim that the laws of B have been reduced to the laws of A, or to $A \cdot L$, if some members of L belong to B; for this would seem to involve an element of what might be called "self-reduction," which is absurd. How can B be reduced to a set of premisses which includes, necessarily, part of B itself?

I suggest that the notion that generates this tangle is misconceived. It is arbitrary and incoherent to permit the "decision" regarding whether or not some member of L belongs to B to depend on whether the biconditional connective or some other truth-functional connective occurs within it. One would think that a question of this sort is (somehow) answered on the basis of the *terms* occurring within a statement rather than on its logical *form*—merely on its truth-functional structure. It seems more plausible to say either that all members of L (including biconditionals) belong to B or that none do. If all, then there is no reduction of B to A. If none, another puzzle arises, for the constituents of L cannot then belong to theory A either, since they contain terms from V_b. The "bridge laws" L seem to have no home in either theory. What are they, then? None of the horns of this trilemma is entirely satisfactory from the point of view of the standard analysis. (I shall show, below, that something resembling the last option *is* acceptable from the perspective of my alternative analysis of reduction. This will constitute one advantage of my analysis.

I would like to emphasize however that Hempel's *motivation* with respect to these issues is quite correct; the first major advantage of using Russell's maxim as an analytical tool rather than the standard analysis will help bring this out. Hempel requires bridge laws of biconditional form because ". . . such a law could then be used to 'define' the [*B*] term, and thus, *theoretically, to avoid it* [my emphasis] ." This is the heart of the matter! It is what Hempel refers to as the "linguistic version" of the "originally ontological view" (making the "conceptual apparatus [of *B*] dispensable"). This is what constitutes the motivation for the type of reduction under consideration. As expressed by scientists motivated to attempt it, this type of reduction is *concept replacement*; it is, also, this aspect of reduction which, in improved form, Hempel wants (quite correctly, in my view) to preserve in his own analysis. Hempel seems to consider an analysis of reduction acceptable only if it captures this pre-analytic aspect of it. He believes that biconditionals *could* be used for such a term or concept *replacement* (avoidance) and that other logical forms could not.

However, biconditionals need not be considered as directing or even suggesting any such replacement. The biconditional is but one truth-functional connective among others. Something stronger is required: *replacement* is an additional step entailing a rereading of the formal notations of the relevant theories—even if all the bridge laws are biconditionals. *It is in fact the crucial step.* (Notice, by the way, that if it were maintained that a biconditional bridge law *is* able to capture the idea of replacement, then so could a *conditional* bridge law which included the relevant term from the theory being reduced within its own antecedent clause.)[7] It is true that the existence of a biconditional bridge law *may* (it need not!) move one to attempt a *replacement*, but the replacement is a separate and rather consequential step not *entailed* by the mere occurrence of the biconditional bridge law in theory *B*.

But if it is the possibility of *replacement* or avoidance (of "entities" or of "terms"—Hempel analyzes the replacement of the former on the basis of replacement of the latter) upon which reducibility of theories ultimately depends, then why not bring this notion of replacement to the forefront of the discussion? Why not focus the analysis upon it rather than link it indirectly with doubtful reasons for favoring one truth-functional form rather than another for the set of bridge laws? My own analysis of concept replacement reduction on the basis of Russell's maxim proposes precisely this kind of reduction, which (exemplified by the electrodynamics/mechanics case) is analyzed primarily as a *replacement of terms* (and, as we shall see in Chapter 4, of the *entities* to which they refer). There are other species of reduction to be considered in later chapters. But it should now be clear why I have called the present type of reduction *concept replacement reduction*.

Advantages of the Concept Replacement Analysis: Concepts vs. Laws

My application of Russell's maxim led to a "concept replacement analysis" of one type of intertheoretic reduction that can be expressed briefly thus:

> One theory *B* has been reduced to another theory *A* if and only if for each *term* of V_b one can construct a function of terms of V_a such that, upon *replacement* of the former by the latter in the *laws* of *B*, one obtains transformed statements which can be shown (in one way or another) to be laws of *A*.

The remainder of this chapter will be devoted to displaying further advantages of this way of analyzing reduction in physical science—advantages due to an emphasis on *concept replacement* rather than *law deducibility*.

I should point out at the outset that there are elements of a correspondence between this and the standard analysis. Some (but not all) of the differences between the two are primarily differences of perspective and emphasis and of what was once called "pragmatics" rather than of (semantic or syntactic) content, though such differences are admittedly important. The elements of the correspondence are these:

a) Where the standard analysis speaks of *bridge laws* of biconditional form for each member of V_b, I speak of *replacement functions* $f_i(V_a)$ for each member of V_b. Bridge laws are stated *within* the formalism common to theories A and B; their subject matter consists of items in the domains of A and B; they are in the "object language." My replacement functions, on the other hand, are *metalinguistic* items; their subject matter consists of *theories A* and B themselves. They are *directives* as to how we should treat the two theories. (This "linguistic difference" is the key to some important conclusions regarding physical ontology, which will be discussed in the next chapter.)

b) Where the standard analysis speaks of *deducing* laws of B from laws of A conjoined with bridge laws L, my analysis speaks of the laws of B *transforming* into laws of A when the replacements for the terms of V_b are made. Of course, through deduction, it could be shown that the transformation is successful. But it *need not* and sometimes *cannot* be thus shown. This is a crucially important difference which, as I have promised, I shall discuss below.

It is not my purpose simply to "refute" the standard analysis. Indeed, it is not *simply* false: certain features of it are preserved in my analysis. The situation is not wholly unlike what takes place when one theory reduces and hence replaces another: the latter is not *simply* rejected; certain structural features are retained. What I shall try to show is that my analysis provides a more coherent way of viewing inhomogeneous theoretic reduction than does the standard analysis. Those who have stayed with me to this point will appreciate that it can be the case, even within physics, that two theories may be formally interderivable though one be better than the other in handling certain sorts of problems. For example, the Hamiltonian and Lagrangian formulations of classical mechanics, though in an appropriate sense *equivalent*, are not *equally appropriate* (in a sense I won't detail here) for certain kinds of physical systems. (Similar remarks may be made regarding the use of different coordinate systems in physics. See, e.g., pp. 4-5 of my book on relativity theory.)

It is in this spirit that I offer my analysis of reduction. Even were it somehow "equivalent" to the standard analysis (which it is *not*),[8] I would argue that it is more appropriate, in a sense still to be clarified. My discussion in the remainder of this chapter will proceed under two main headings, but the considerations offered will overlap.

(1) Term replacement vs. law deducibility as central

The standard analysis likens reduction to explanation. Indeed, Nagel says that *all* reductions *are* explanations: "Whatever else may be said about reductions in science, it is safe to say they are commonly taken to be explanations, and I will so regard them."[9] But explanation, according to a related "standard view" (the deductive-nomological analysis associated with Hempel,[10] to which Nagel has subscribed), consists of a set of statements arranged in a particular *deductive* way: reductions of theories are there viewed primarily as deductions of the statements (laws) of one theory, B from the laws of the other, A. Attention is thus immediately and primarily focused upon the *laws* of the relevant theories and their *deductive* relations. A "problem" is said to arise with such deductions of laws if some of the *terms* of B do not occur in A; for such ("heterogeneous") reductions, the second and third elements of the analysis are introduced, that is, the requisite *bridge laws* to relate the two sets of *terms*.

I will have more to say about the relation between explanation and reduction later; for the moment, we need note only that, even *if* the standard analysis is not in any simple sense "incorrect," it presents the issues in a distorted way. For, in at least some cases, the scientific motivation that leads to attempted reductions does not directly involve *deduction* of laws but rather *replacement of terms—concepts* and the *entities* (things, properties, relations, processes, etc.) that are their purported designata (Russell's "inferred entities," indeed). That is, at least some attempted reductions in science are better described (in the terminology of the standard analysis) as attempts to find bridge laws for specific bothersome terms or concepts. The deductions of the relevant laws are then considered as a *test* of the correctness of the proposed bridge laws. At least, this is what would have to be said within the framework of the standard analysis in order better to reflect what scientists are doing in at least some cases of reduction. Although the apparatus of the standard analysis does allow one to say this much (I've just said it), the remark would properly occur as a non-essential, pragmatic fact about reduction; reduction itself would have to be *analyzed* in accord with conditions (i) and (ii), provided earlier. The point is more naturally accounted for in my concept replacement analysis because it is contained in the very statement of the analysis rather than appended as a comment regarding the pragmatics of reduction. The difference is not trivial, for reductions of the kind under consideration are *directly described as attempts to find replacements* (replacement functions) for the concepts of the theory being reduced. These replacement functions must meet the general condition that they transform laws of B into laws of A. In my analysis, condition (i) of the standard analysis thus becomes but *one* kind of *test* (not even a necessary test) of the success of the attempted replacement of concepts: given these new statements couched in the terminology of theory A (which have been generated by the systematic replacement of B-terms by constructs of A-terms in the laws of theory B), can they be shown to be laws of A *in some way or other*—by deduction, say, from known laws of theory A? (We have, here, an example of the manner in which my analysis partly preserves an element of the structure of the standard analysis.)

One could in fact go further and say that theoretic reduction of the type under consideration *is* the replacement of concepts—*tout court*. In a sense reminiscent of Michael Scriven's view of the role of deduction in explanation,[11] the *deducibility* of laws would *not* be considered *part* of the reduction itself but rather part of a possible *justification* that might be given for the correctness of the reduction. Even were this an exaggeration, it affords a more illuminating view of reduction than does the standard analysis. (Of course, it must be shown, in *some* way, that the laws of B are transformed into laws of A.) I will have more to say about this in section 2 below.

Those not convinced by my earlier arguments relating to the nature of bridge laws may still believe that there is no difference in *formal* content between the two descriptions of reduction. They may see the difference as merely methodological or "pragmatic" (rather than syntactic or semantic). But I trust that few will deny that the ability to capture just these methodological or pragmatic features of science and scientific procedure—the reductive *enterprise* itself—is a philosophical gain.

Attempts during the latter half of the nineteenth century to construct "models of the ether"—to reduce electrodynamics to the remainder of mechanics—provide examples of reductive attempts most naturally viewed as *replacements* of *terms* or *concepts* (or of purported *entities*). (See for example the long quotation from Whittaker, in the previous chapter, where the burden of the description of the attempted reduction *is* the suggested concept replacement.) During this period, investigators often begin by offering such replacement proposals. Further deductions are then construed as attempts to verify the correctness of the original concept replacement proposals. One does *not* find physicists thus engaged *primarily* attempting to deduce electrodynamic laws from mechanical laws (although they of course do attempt this) or constructing "bridge laws" as they are needed to effect deductions or stumbling upon likely candidates as they proceed. They wish instead to show, for instance, that "electric fields really are strains in the ether."

Essentially the same point has been made by other authors who offer accounts of reduction different from mine and that of the standard analysis. For example, almost in passing, Ager, Aronson, and Weingard remark:

> In 1924, Louis Victor de Broglie proposed that all particle phenomena can be treated as wave phenomena. This is clearly a reductionist thesis. But contrary to the orthodox view of reduction [the standard analysis], de Broglie's problem was not to derive the laws of particle mechanics from the laws of wave mechanics. Rather the problem which immediately arose was whether certain particle properties were unambiguously identifiable with wave properties.[12]

The idealized statements of philosophers of science regarding the proper analysis of concepts in science are seldom displayed in "pure form" in the actual writings or activities of scientists. But I would like to offer one final example of what I take to be a relatively clear case of the better fit that the concept replacement analysis provides for the kind of reduction under consideration. Here is a somewhat lengthy quotation from a scientist regarding the relation between thermodynamics and statistical mechanics (a paradigm case of reduction and a workhorse example used by philosophers):

> . . . *thermodynamic* questions can be answered in terms of the underlying micro-description. How? The state variables appropriate to the system at the *gas* level [*B*-terms] must be expressed in terms of the observables [not the philosopher's sense of this term] of the underlying mechanical system; i.e., as functions of the state variables of the microsystem [*A*-terms]. It is then a question of finding the appropriate observables and the operations which must be performed on them. This is what statistical mechanics does. It identifies a thermodynamic state (macrostate) with a class of underlying microstates, and then expresses the thermodynamic state variables as averages of appropriately chosen micro-observables over the corresponding class of microstates. The equations of motion of the underlying microsystem [the laws of *A*] are then inherited by the thermodynamic state variables, allowing us to express in principle (though in practice only as yet in very special situations) the kinetic behavior of the system at the macrolevel [the laws of *B*].[13]

Notice that we don't even know, except "in principle," that the laws of *B* in this case are deducible from ("inherited by") those of *A*, but it is still firmly believed that this is a case of a successful reduction because it is believed that we have constructed the appropriate replacement functions for terms. No one doubts that gases are composed of molecules in motion, even without full deducibility of laws. We have other ways to justify the replacement—other ways to test the hypotheses regarding the microsystems involved besides the deducibility of the gas laws of thermodynamics.

So it is *not* the case that deduction of laws is primary—with so-called bridge laws required to accomplish this; rather, *first*, replacement functions are stated and *then*, one tries to determine the relevant laws in at least certain simple cases ("very special situations"). Deduction of laws, therefore, is not a necessary condition for successful theory reduction. The next section will include an elaboration of this point.

(2) Law deduction not even necessary for theory reduction

The emphasis on replacement of terms rather than deduction of laws has a further consequence. There are, actually, forms of reasoning favored in scientific reduction that could not be accommodated if the deduction of laws were treated as a necessary condition of reduction (condition [i]); they can, however, be described very naturally in terms of the term-replacement analysis. (In such situations, the standard analysis is simply regarded as incorrect.)

Sometimes, upon replacement of all B-terms by a proffered set of (complex) functions of A-terms in the laws of theory B, one will obtain very complex new law-like statements couched in the language of A. Call one of them 'C'. It may be the case that C is too complex to be *deduced* from already known laws (or postulates) of A. Condition (i) of the standard analysis is thus not satisfied. Researchers will, instead, attempt to *verify* C—to show it to be true—*in some other way*. They may for example simply try to compare C directly with experience in the domain of A. Or they may compare consequences of C with experience in the domain of A. Such methods failing (the experiments involved may be too complex to be carried out, the effects too small to detect, etc.), they may try the following rather interesting procedure: Deduce consequence C' from C (where C' may be a rather remote consequence); then, translate C' back into the terminology of theory B via the replacement functions used in reverse; finally, directly compare this retranslated deduction (or consequences of it) with experimental results in the *domain of theory B*.

Now, it will be considered good evidence that the original replacement functions are correct—that the reduction of theory B to theory A is successful—if the relevant observations are in accord with the test statement, by way of any of the above procedures (or combinations of them). (The last procedure described is a way in which "new effects" are found in the domain of the theory being reduced; for often, the retranslated C' will describe previously unexpected type of behavior. It will provide particularly convincing evidence for C and thus for the replacement functions which led to C. Therefore, it provides evidence for the correctness of the reduction, even though theory A is used only as a temporary bridge to obtain further predictions in the domain of theory B.) Viewing such operations from the standpoint of the standard analysis, one would have to conclude that reduction has *not* taken place; after all, condition (i) has simply not been fulfilled. In each case, *C* (perhaps other law-like statements of the same sort as well) has *not been deduced* from already known laws of *A*—the postulates of *A* in particular.

But according to my concept replacement analysis, it is only necessary that the laws of *B* be transformed into laws of *A* (via the replacement functions). Admittedly, *one way* of telling whether *C* is a law of theory *A* is to deduce *C* from *A*'s postulates or from other laws of *A*, but *it is not the only way*. Other ways include what I have just described: directly comparing it, or consequences of it, or various retranslated consequences of it, *with experience. A true law-like statement couched in the terminology of theory A is a law of A, regardless of how its truth is established.*[14] It is a well-known fact of scientific life (usually overlooked by philosophers of science) that at any given time in a field of investigation there are *known laws* which have not (yet) been deduced from "the" basic laws of the field. Such a deduction is *not* requisite for *knowing* them to be *laws*.

In this respect, science is full of things resembling "Fermat's Last Theorem" (if *n* is a number greater than two, there are no whole numbers, *a*, *b*, *c*, such that $a^n + b^n = c^n$) or "Goldbach's Theorem" (every even number is the sum of two primes). Regarding the latter, it has been remarked that "It is easy to understand; and *there is every reason to believe that it is true* [my emphasis], no even number having ever been found which is *not* the sum of two primes; yet, no one has succeeded in finding a proof [deduction] valid for all even numbers."[15] Now I am certainly *not* about to argue, on the basis of this quotation, that "valid proofs" in the form of *deductions* are not required in *mathematics*. (Although such a case can, and has, been made![16]) But I do want it to be clearly understood that in (empirical) *science*, at least, there are other ways of establishing laws besides deduction from other more basic laws; and that, often, these other ways are all that we have, though they are also sufficient for confirming a successful reduction.

An example of reducibility without deducibility is provided by "tide theory," which no one would deny is reducible to mechanics—Newton's great unification accomplished this. Yet, specific tidal regularities (for particular places) simply cannot be deduced from basic mechanical laws (plus boundary conditions). The complexity involved is simply over-whelming. Another example is provided by that sub-domain of me-chanics known as celestial mechanics. No one doubts that this is part of mechanics. Yet the systems involved are too complex to permit deducing detailed observed behavior from basic mechanical laws. Certainly, before the advent of modern computers, there were many observed regularities involving planctary motions that could not be calculated on the basis of (deduced from) $F = ma$ plus the law of uni-versal gravitation (plus boundary and initial conditions). Yet celestial phenomena of this sort were still (properly) considered to have been *reduced to* mechanics. After all, this was another part of Newton's great unification.[17]

It might be said in response that the laws of B in each of these cases *are* deducible from those of A; the deductions have not yet been actually achieved (due to a lack of computer time, for example). But this response will not do, for if it is *believed* that all the laws of B are deducible from A (even though many have not been deduced), the be-lief is due to the prior belief that the reduction has already succeeded. Therefore, the deducibility condition cannot be used as a necessary condition to determine *whether* a reduction has been achieved. If a reduction (in my sense) has been achieved, then scientists may well be prompted to attempt law deductions not previously attempted.

Other such cases could be mentioned,[18] but I believe the point is sufficiently clear. Still, it may be instructive to examine briefly what lies behind the deducibility requirement of the standard analysis. Why do those who defend the standard analysis require law deducibility in the first place?

If intertheoretic reduction is characterized as one kind of *explanation*, the difficulties confronting the standard analysis in the situations I have been describing become almost inevitable. For proponents of the standard analysis, explanation of a theory is explanation of its *laws*; and explanation of laws is analyzed as *deduction* of them. A failure to de-duce a given law means a failure to explain it; and that, in turn, means a failure to reduce it as well.

But if my arguments carry weight, we need not infer the non-reduc-tion of a law on the strength of its non-deduction. (I will have more to say about this complex issue, in Chapter 4.) Alternatively, if a law has not been reduced, ought we to conclude that the relevant *theory* cannot have been reduced? This is claimed also by the standard analysis, since a theory is usually *identified*, on that view, with a *conjunction of (its) laws*—not merely with all of its laws but with its "basic laws" and with whatever is deducible from them. I do not want to go into a lengthy discussion, here, of the nature of a *scientific theory*, but I suggest that it would be too narrow a construal of the concept to identify a theory with (some given set of) *its laws*. For one thing, it disallows cases of reduction of the sort I have been describing. The standard analysis could avoid precluding such reduction only by denying that theories are to be identified with conjunctions of (some of their) laws.

If theories were identified with conjunctions of *all* of their laws, then the standard analysis could not require deduction of laws as a condition of reduction of theories, for this would amount to requiring that any theory which is to serve to reduce another must be deductively complete. This assumes that theories are deductively unified to a degree not found in areas of science where reductions nevertheless do take place. Examples have already been provided. (Indeed, we saw that they appear in mathematics as well.) This version of the standard analysis, then, implicitly construes science as consisting wholly of interpreted *formal systems*. Such an analysis may well be an illuminating way of viewing *certain aspects* of the nature of science, but it ought not to be used as a procrustean bed.

Any acceptable analysis of the nature of scientific theories must include, centrally, a view as to the basic kinds of *entities* (things, properties, . . .) constitutive of, and operative in, the domain in question. Scientific theories make claims about what there is. (Much more will be said about ontological issues, in Chapter 4. For an example of the present point, see Chapter 4.) Its inclusion would permit the *reduction* of one theory to another to be described as the replacement of *terms* designating *entities* referred to. The *establishment* of laws then appears as a (necessary) condition for *certifying* any *term replacement* as correct, and the *deduction* of laws appears as but one way to satisfy this condition. The use of *this* particular way of certifying that a certain term replacement is correct serves as a measure of the deductive unity of the reducing theory—by no means a trivial consideration; but it is not a measure of *whether* the reduced theory has in fact been reduced. It may, of course, provide further evidence that the reduction has indeed been accomplished; such evidence is never unwelcome, but it is not always *required*. We may, for instance, already have overwhelming evidence that a theory reduction has been accomplished—perhaps through the direct empirical verification of the laws in question (not necessarily of all of these). Further deductions might facilitate a certain "mop-up work," without affecting the already settled question of the reduction's success.

I may put the point briefly (perhaps too briefly) as follows: Laws describe how the entities of the domain of a theory behave. Theory reduction (of the type under consideration) is *entity reduction*, as expressed by successful replacement functions for terms. (Again, more will be said about ontological issues, in Chapter 4.) Deduction is one (good) way of establishing laws, but not the only way. Nor does this detract from the other functions which the deduction of laws may serve. (It should now be clearer why Scriven's analysis of explanation, to which I referred earlier, provides a model for understanding what I am attempting to say about the relation between *theoretic reduction* and *deduction of laws*.)

I would like, now, to consider briefly two kinds of response to my analysis that a defender of the standard analysis might offer.

The first is that my distinction between what is part of a reduction and what is only part of the justification for a claim that a reduction has been accomplished is only a quibble; hence, that my view of the deduction of laws as a possible part of the latter rather than a necessary part of the former is neither interesting nor an important departure from the standard analysis. Several replies may be made. First, does my way of viewing this type of reduction reflect scientific practice more accurately? I believe it does, whether or not this aspect of scientific practice is judged important or interesting. The deduction of laws usually serves as (but one method among others for the) certification that a concept replacement reduction has taken place rather than as a necessary part of the claimed reduction itself. Secondly, treating the deduction of laws as a possible way of justifying reduction is necessary, since there are recognized reductions where law deduction is simply not fulfilled. But this brings me to the second kind of response that may be offered by a defender of the standard analysis. A defender might maintain that, notwithstanding scientific practice, one *must* (or *ought to*) have achieved a deduction of laws before *validly* claiming to have achieved a reduction. The deduction of laws—even if construed to be part of the justification for a reductive claim rather than part of that reduction itself—must be at least a *necessary part* of any genuine justification. If this were a claim about actual scientific practice, it would simply be false. It could, however, be more reasonably interpreted as a normative claim. It would then serve as an exhortation to scientists not to claim success until the law deducibility condition were met. Thus interpreted, the standard analysis says that science should be practiced in a way different from that which now obtains. The question remains, what advantages or disadvantages would there be if science were to be practiced in that way?

As a matter of fact, there would be no difference regarding the effort to deduce laws, because the deduction of laws would be energetically pursued *in its own right*, whether or not such deductions are considered to be part of any theory reduction. A scientist may well claim a successful reduction without having achieved law deductions; he may, of course, also attempt to achieve law deductions because he is *also* interested in the deductive unification of the deeper theory. This, perhaps, accounts for the plausibility of including the deduction of laws as part of the analysis of the reduction of theories. (But though it may be true that X is a goal worth pursuing as part of endeavor E_1, it does not follow that it ought to be pursued as part of another endeavor, E_2.)

In conclusion, I believe I have achieved two results. First, I have indicated a number of difficulties in the standard analysis of reduction avoided by my concept replacement analysis. Second, my analysis places the role of laws and law deducibility in its proper perspective in relation to the role of term or concept replacement.

Notes

1. See for example Hempel's "Reduction: Ontological and Linguistic Facets," in S. Morgenbesser, P. Suppes, and M. White (eds.), *Philosophy, Science, and Method; Essays in Honor of Ernest Nagel*, New York, 1969, pp. 179-199. Nagel's most recent published views on reduction appear in his "Issues in the Logic of Reductive Explanations," in H. Kiefer and M. Munitz (eds.), *Mind, Science, and History*, Albany, 1970, pp. 117-137. Each of these papers contains references to the earlier works of both authors on the nature of reduction as well as to the work of others. In particular, see Nagel's "The Meaning of Reduction in the

Natural Sciences," in A. Danto and S. Morgenbesser (eds.), *Philosophy of Science*, New York, 1960, pp. 288-312 (originally published in 1949); and Nagel's book, *The Structure of Science*, New York, 1961, Chapter 11.

2. *If* they are different. If V_b contains no terms which do not also occur in V_a, then this second condition is trivially fulfilled and we have what Nagel calls a "homogeneous" reduction. I am primarily interested, therefore, in what he has called "*in*homogeneous" reductions.

3. Not all presentations of the standard analysis state the condition so strongly. But it will become apparent as we proceed that something at least this strong is required.

4. This is from Hempel's "Reduction: Ontological and Linguistic Facets," *op. cit.*, p. 189. My bracketed insertions involve removal of Hempel's specific example, biology and physics, in favor of a more general statement of the position.

5. Bridge laws are often of conditional form, providing a necessary or sufficient condition (but not both) for the application of a term from V_b on the basis of terms from V_a.

6. Consider the following miniature case: Theory A contains two postulates, $Q_1 \supset Q_2$, and $Q_2 \supset Q_3$; and theory B contains one postulate, $R_1 \supset R_2$. The set L of bridge laws consists of two members, $R_1 \supset Q_1$ and $Q_3 \supset R_2$. Condition (i) of the standard analysis is met for this case, because "all" of theory B's postulates are deducible from the conjunction of L with the postulates of theory A. But condition (ii) is not met, because there are no biconditional bridge laws for the terms of V_b (the terms 'R_1' and 'R_2'). This demonstrates that any assumptions that imply that satisfaction of condition (i) *entails* partial satisfaction of condition (ii) must be in error.

7. A few more words on this point may be in order. Hempel argues that *non*-biconditional bridge laws are not sufficient for claiming a successful reduction, for a non-biconditional bridge law is merely an additional law of B, the reduced theory (or the theory we are attempting to reduce). Such laws would "simply expand the theory B." Now if '$a \supset b$' (say, where 'a' and 'b' represent functions of terms from V_a and V_b respectively) is a law of B, then so of course must be '$b \supset a$' (if it is accepted as true). But then, their conjunction, '$(a \supset b) \cdot (b \supset a)$' must also be a law of B (or at least "part" of B if the notion of a "conjunctive law" sounds odd). But this last "law" is equivalent to the biconditional '$a \equiv b$', and therefore this biconditional bridge law must also be a *law of B*.

The essential point is that truth-functional connectives are *just that*: ways of constructing compound statements out of simpler statements in a manner in which the truth value of the compound is wholly determined by the truth values of the components. In this sense, '$\equiv$' is in the same boat as '$\supset$' (or as '$\mathbf{v}$' or '$\cdot$').

8. The two might *appear* to be equivalent in this sense: For any two theories B and A, B is reducible to A according to the standard analysis if and only if B is reducible to A according to my concept replacement analysis (leaving aside for the moment the problem of which theory L may be said to belong to). But even this sort of weak equivalence does *not* hold if the considerations I shall offer in section 2 below have merit. (Moreover, we have already seen difficulties in the standard analysis which do not arise for my concept replacement analysis.)

9. "Issues in the Logic of Reductive Explanations," *op. cit.*, p. 119. The very title of the paper makes the point. The claim is also implicit in the quotation from Larmor in note 13 of Chapter 2, above, where Larmor speaks of the introduction of "concealed structure" as an *explanatory* attempt. (I shall say more later on the relation between reduction and explanation.)

10. For a full description of this analysis of explanation see Hempel's "Aspects of Scientific Explanation," in a book by the same title and author (New York, 1965).

11. Michael Scriven, "Explanations, Predictions, and Laws," in H. Feigl and G. Maxwell (eds.), *Minnesota Studies in the Philosophy of Science*, Vol. 3, Minneapolis, 1962. See especially pp. 196 ff. The relevant parts of this paper are also reprinted in B. Brody (ed.), *Readings in the Philosophy of Science*, Englewood Cliffs, 1970.

12. Tryg A. Ager, Jerrold L. Aronson, and Robert Weingard, "Are Bridge Laws Really Necessary?," *Nous*, Vol. 8, No. 2, May, 1974, p. 131.

13. Robert Rosen, "Hierarchical Organization in Automata Theoretic Models of Biological Systems," in Whyte, Wilson, and Wilson (eds.), *Hierarchical Structures*, New York, 1969, p. 186.

14. It would be sufficient to consult a reputable oracle, or a "confirmation machine." See Edward Erwin, "The Confirmation Machine," in R. Buck and R. S. Cohen (eds.), *Boston Studies in the Philosophy of Science*, Vol. 8, Dordrecht-Holland, 1971, pp. 306-321.

15. E. Kasner and J. Newman, *Mathematics and the Imagination*, New York, 1940, p. 187.

16. See Mark Steiner, *Mathematical Knowledge*, Ithaca, 1975, Chapter 3: "Proof and Mathematical Knowledge." Steiner argues for the existence of inductive evidence as justifying knowledge claims in some mathematical cases in which a deductive proof cannot be constructed.

17. I am here open to the accusation that I am confusing two relations: (1) the relation of reducibility that holds between two theories, and (2) the "branch" relation that holds between a theory and one of its areas of application. But I believe there to be no significant difference here. Once a concept replacement reduction of theory B to theory A has been achieved, then a (perhaps reinterpreted) B may be viewed as a branch of A. This is of great significance; I will have more to say about this kind of situation, in Chapter 8.

18. See the chemistry/physics example of the reduction of one *branch of science* to another, in Chapter 7.

Chapter 4

Concept Replacement vs.
the Standard Analysis of Reduction—Part II

This chapter will be concerned with certain further aspects of my concept replacement analysis of one kind of intertheoretic reduction. In particular, the focus will be on the idea of *replacement*. I shall argue that my analysis has implications of significance in two areas: physical *ontology* and the concept of a scientific *explanation*. I begin by discussing an important advantage of interpreting the relationship of terms of V_a and V_b on the basis of *metatheoretic* replacement functions rather than of "bridge laws" formulated in the object language common to a pair of theories.

Metatheoretic Replacement vs. Object Language Bridge Laws

If bridge laws are physical *laws*, then *they invite explanation* (even if they are considered postulates they eventually require explanation in terms of some future deeper theory). But if explanation requires deduction from more basic laws, then the set of bridge laws L relating the vocabularies of theories A and B should be deducible from the postulates of the reducing theory A, since it is the more basic theory. But the members of L cannot be deduced from the postulates of A, for, by hypothesis, they contain terms which are not in A's vocabulary V_a. The set L was introduced of course for the very purpose of connecting the vocabularies of A and B. (This relates to the difficulties of the standard analysis discussed earlier, in Chapter 3.) This puts the set of bridge laws L in an awkward position: the members of L

do not belong to A but appear to belong to B. But I have shown (in Chapter 3) that the assumption that they belong to B leads to a regress, if we assume that B has been reduced to A *with the aid of L*.

The members of L, therefore, must *belong to neither theory* (as I argued in Chapter 3); moreover, they are not in any sense "basic" but seem to require an explanation—which they cannot receive. The members of L actually resemble the "nomological danglers" of which Herbert Feigl has written in another context.[1] We have before us a problem similar to one which he raised for mind-body dualism.

My alternative analysis, on the other hand, *begins* by maintaining that they do not belong to either theory A or theory B. The "items" which we are considering are *not physical laws at all*. They do not state or express any pattern of interaction or association among distinct entities in the universe. They do not describe physical regularities which would require an account on the basis of deeper laws or regularities regarding the interaction of objects in the physical universe. Rather than *belonging to* any theory, they express relations holding *between* two of our theories—relations which cannot and do not need to be deduced from either of the theories being related (or from any third theory).[2] They are *metalinguistic*.

Others have reached a similar conclusion—that the "items" which connect the vocabularies V_a and V_b do not require explanation; but they have done so by maintaining that these items (so-called "bridge laws") should be construed as *identities* rather than biconditionals and that contingent identity statements do not require and cannot receive an explanation.[3] The issue of whether identity statements require explanation has by no means been settled; it has in fact been argued that identities *can* be explained.[4] It is thus not an inconsequential advantage of my *replacement* analysis, that I am able to reach the important conclusion that the manner in which V_a and V_b are related is not something requiring an explanation, without appealing to arguments about the nature of contingent identity statements or about whether or not they require or can receive an explanation.

Let me make the present point clearer by way of some examples. It will be useful to begin by going back to the application of Russell's maxim in mathematics. Consider the case of number theory, in which Russell's maxim bids us to replace irrational number expressions in the theory of irrational numbers by rational number expressions. No mathematician would consider it appropriate to ask for a *deduction* (explanation) of the statement of the replacement proposal from the theory of rational numbers (or from any other mathematical theory). Suppose however that we were to reread these metatheoretic replacement proposals in the manner of object language "bridge laws" (*or* identities). We would then have a set (an infinite set) of mathematical statements couched in the language common to the theories of rational and irrational numbers, in which a correspondence was drawn between rational and irrational numbers. These would generate such questions as, "Why are these true?" and "Can we deduce them from higher-level mathematical theorems?"—utterly inappropriate questions.

Consider also examples from metaphysics. Instead of interpreting physicalism as the claim that we may *replace* theoretical object talk or theory by ordinary physical object talk or "theory," we would have to consider instead the claim that theoretical objects and physical objects display various joint regularities, each one of which would invite an explanation (on the basis of what?). But the physicalist program does not involve the claim (e.g.) that electrons are present if and only if ammeters are behaving in a certain way.[5] For that claim would generate the question "Why is this so—in accordance with what deeper laws do electrons and ammeters interact in this way?" That question is coherent only as a *realist* question *within* physics. The *physicalist* claim is rather the *metalinguistic* one that theoretical object language (physics itself) is to be *replaced by* talk about molar objects such as ammeters. There is no regularity at issue to be explained. Similar remarks apply to phenomenalist programs. These do not claim law-like regularities holding between ordinary physical objects on the one hand and sense-data on the other; rather they claim that one way of describing the world may be *replaced* by the other. Any alleged *regularities* holding between physical objects and sense-data could make sense only (if at all) as statements within the psychology of perception interpreted in a *realist* way. (Incidentally, reductive *motivations* in such cases are surely *ontological* rather than linguistic. But reductive programs are often carried out in linguistic garb, and I am at the moment comparing them in a linguistic way. I will have more to say about ontological aspects later in this chapter.)

This is the kind of claim I am making regarding our examples of intertheoretic reduction *within physics*: the question (e.g.) "Why are electric fields uniformly associated with strains in the ether?" (assuming that this reduction had succeeded) is not an appropriate question, as it would have been if the relation in question were interpreted in terms of a "bridge law"; for no physical regularity calling for explanation has actually been stated. Rather, there are two theories—two ways of describing certain aspects of the universe— and the claim is being made that one may be *replaced* by the other (via a replacement of the terms of electrodynamics by complexes of terms from the remainder of mechanics). Replacement functions express the details of this relation between the theories in question.

Briefly put: bridge laws express relations among *physical entities* described by our theories about the world, while replacement functions express relations between our *characterizations* of the world. As such, questions appropriate to one are not appropriate to the other—which resolves the problem of nomological danglers.

I hasten to add that explanations may be invited *regarding* replacement functions (though not *of* them). For example, it may reasonably be asked in a particular case: "Why wasn't the replacement possibility noticed earlier?" It might for instance be asked why it was not noticed earlier that the electric field concept could be replaced by the concept of a mechanical strain in an underlying elastic medium. Answers to such questions are contained only partly *in* the theories in question. They require information as well about *our* relation (as investigators) to these theories, their domains, and the instrumentation used by us to apply the theories to their domains. Perhaps we had not noticed the possibility earlier because of the particular turn that investigation and observation had taken; perhaps this, too, could be explained on the grounds that the posited strains in the mechanical ether were more difficult to treat observationally in their detailed structure than certain of their averaged-out molar features. If people were as small as atoms (but otherwise pretty much the same as they are now), electric field language and ontology might never have been introduced in the first place—or, introduced later only as shorthand for already known complex facts about strains in the mechanical ether.

From another perspective, I have been claiming that a successful replacement program—a successful reduction—amounts to showing that there is but one set of phenomena, one domain, in question, describable in two different ways. Previously, it was believed that one such description was wholly inappropriate to the phenomenon in question; or, its appropriateness may not even have been considered. This raises some further ontological issues. But before I discuss them a few preliminary remarks may be useful.

A few paragraphs back, I described two metaphysical views—physicalism and phenomenalism—as claims about *languages* or *theories* which can (or ought to) be used in describing the world. Historically, of course, these are *ontologies* (or at least *ontologically motivated* views about language); they constitute positions regarding what there is. In his paper "Relation of Sense-Data to Physics," Russell for example did not think of himself as having made claims *merely* about how to *talk* about the world; rather, he thought he had put forward a view about what, ultimately, we must assume *exists* in the world. In this sense, descriptions of these metaphysical positions may be viewed as the linguistic form of certain ontological claims. Similarly, my analysis of one kind of theoretical reduction in physical science was formulated as if it were a claim about ways of *talking* about the world: I said for instance that the standard analysis's "object language" bridge laws ought to be reconstrued as "metalinguistic" replacement functions. But that analysis can be understood as well as maintaining that theory reductions involve *ontological* claims and implications. From this perspective, my claim about the "metalinguistic nature" of "bridge laws" is merely linguistic clothing for what is at bottom an ontological claim.

Many philosophers today regard the *analysis* of ontological claims as an analysis of claims about language. This is Hempel's view, for example, favored in the paper discussed earlier. Those who find this point of view congenial may construe my account conformably; indeed, I have stated the anlaysis that way. Those who feel otherwise about ontology may consider my analysis as making claims about physical ontology. What I have said thus far can be read in either way; I take this to be a mark of the strength of my analysis.

However one resolves the *philosophical* question posed, when *scientists* consider specific reductions within physical science, such reductions are regularly viewed as (what philosophers would call) *ontological* in nature. What Hempel calls "the linguistic construal of reduction" (*op. cit.*, p. 180) is a creature of the middle of this century; and, as even he admits:

. . . one further reason for the fascination the subject [reduction] has held for philosophers lies, I think, in the ontological roots of many questions concerning reduction. (*op. cit.*, p. 179)

Whether or not it is fully analyzable in a linguistic manner, physical *ontology* lies at the center of the reductive enterprise, for physical theories do purport to provide an account of what there is. In the next section, I shall consider one particular ontological issue regarding my use of metatheoretic replacement functions rather than object language bridge laws (or identities).

Replacement vs. Identification: Some Remarks about Ontology

Despite the difficulties of the identity interpretation of the relation between V_a and V_b, it seems enticing enough to make one wonder whether it can be used in conjunction with my replacement analysis. Let us consider the possibility.

The proposal would involve maintaining that there is an (empirically discovered) *identity* between the corresponding referents of the reduced and reducing theories, that is, that there is *one* phenomenon and two correct descriptions of it. My *replacement* functions would then be construed as proposals to *identify* the referent of each term from theory B with the corresponding referent of some complex expression from theory A. (We would simply have to ignore the type of problem raised by Thomson's criticism, *loc. cit.*; and we would have to *assume* that identities do not require explanation.) Such an "identification proposal" would be successful if the laws of B—their terms appropriately *replaced* (this still takes place!) in accordance with the stated identities—would become accepted laws of A. One would show that this could be done by adducing evidence of one sort or another including, though not requiring, deduction of these laws. (See Chapter 3 for various other ways of obtaining such evidence.)

Leaving aside the difficulties of the identity analysis already pointed out, such an amalgamation will not do. For an identification of the referents of theory B with appropriate referents of theory A entails that the terms of B do have referents. I have, however, been arguing that reduction of B to A is best understood as *replacement* of theory B by theory A—replacement of one theory *and its ontology* by another.

Yet, although an amalgamation of the identity and replacement views will not work, the two are not in conflict *if* properly understood: they simply describe two different, complementary attitudes toward a successful reduction.

To show this, let me distinguish two attitudes that one may adopt toward a physical theory. The first, the *theoretical attitude*, focuses on physical *ontology*. It offers a perspective from which a theory is primarily viewed as descriptive of whatever entities, structures, processes, etc. exist at bottom in its domain. Its *basic* laws are viewed as descriptive of the (perhaps relatively unobservable) *fundamental* entities, structures, and processes constitutive of that portion of physical reality with which it deals. Its *derivative* laws are viewed as descriptive of the more complex structures and processes which the fundamental items are capable of entering into in accordance with the fundamental laws. These more complex levels of structure and interaction—*organization*—are not viewed as introducing anything new into the *ontology* of the theory, which is fully determined on the basis of the fundamental laws. The other attitude, the *practical attitude* pays more attention to the *application* of the theory and related uses. The theory is primarily viewed as a *means* for predicting, correlating, and controlling the more or less observable phenomena in its domain of application—a conceptual tool for finding our way around among the observed phenomena or, for example an aid in building radios and reactors or for sending vehicles to the moon and beyond. Notice that the practical attitude is not restricted to technological applications. For example, one who views the kinetic theory of gases as a tool for accurately determining macroscopic gas diffusion laws is operating in accord with the practical attitude. This last concern is of course "theoretical" in a common meaning of the term. But it is *application* that counts, whether technological or "theoretical," rather than the analysis of physical *ontology*.

For present purposes, we note only that the two attitudes are not mutually exclusive. A physical theory may be viewed from either perspective.[6] (It may be added that, although this is true, not all theories function equally in both respects. For example, since it is known to be basically a false theory, classical mechanics is not used for the theoretical purposes I have been discussing; yet it is very useful from the practical point of view. Another type of reduction is involved here, to be discussed in Chapter 5.) I trust that these distinctions may be regarded as merely terminological.

Now, is reduction to be understood as *replacement* of the ontology of *B* with that of *A*; or is it to be construed as the *identification* of the two? Briefly, the answer is this: from what I have called the *theoretical* point of view, reduction is best construed as *replacement of ontology*, for the reasons given. From the *practical* point of view, reduction *is* often—perhaps is almost always—construed as identification (though sometimes as replacement). Let me expand on these ideas.

If a proposed reduction is successful in the sense discussed earlier, then *from the theoretical point of view there is no further need for the reduced theory B.* The ontology of the reducing theory *A* is fully adequate (or at least as adequate as *B* was) for describing what is taking place physically in the original domain of *B*. For example, if the attempt to reduce electrodynamics to the remainder of mechanics had succeeded, we could now speak of mechanical strains in the elastic ether and travelling mechanical waves therein *instead of* electric fields and electromagnetic waves. We could express this forcefully by saying that electric fields do not exist after all, meaning that the ontology of the remainder of mechanics has been discovered to be fully adequate for the theoretical purpose of describing what exists and what takes place in the old domain of electrodynamics (now understood as complex mechanical structures and processes governed by mechanical laws). *Further distinct entities need no longer be assumed.* Using Russell's terminology, "Constructions out of known entities [from *A*]" have been "*substituted*" for "unknown entities [from *B*]."

However, from the *practical* point of view, the reduced theory may well be indispensable for other important purposes or, if not actually indispensable, particularly useful. For example, the theory may be easier or more convenient to apply, may use simpler mathematics, may be more familiar or easier to grasp. We may, for quite good reasons, wish to continue using the apparatus of *B* (conceptual, notational, instrumental) for *predicting* phenomena, *building* and *repairing* artifacts, *explaining* phenomena to tyros in the field, or the like. (As already remarked, such uses, though "practical" in the sense given, may also be called "theoretical" on other grounds. We may for instance wish to predict phenomena which are of "theoretical" (rather than technological) significance. One very intriguing "theoretical application" will be discussed below.)

When involved in *practical* concerns, we *keep* and use the language of *B* and often speak and perhaps even think in ways which belie our *theoretical* knowledge that *B* has been reduced to and replaced by another more basic theory *A*. We often do so (correctly) when the underlying details represented by the reducing theory are not relevant to the concerns at hand. From the perspective of the reducing theory, the reduced theory represents a relatively stable level of structure and interaction; it merely becomes convenient sometimes to ignore the underlying details and to refer to these stable structures and processes as if they were independent entities. In particular, we often speak in a manner that suggests our acceptance of the ontology of the reduced theory *B*.

We may, then, pause from time to time to pay homage to the fact that *B* has indeed been reduced to a deeper theory *A*, "admitting" that the entities of which we speak "are really" only complexes of entities in the ontology of theory *A* (or, ". . . are nothing but . . . ," ". . . are really the same as . . . ," ". . . are *identical* with . . . ," those entities). Thus, we may eat our ontological cake and have it too;[7] we may assert that the referents of the theory *B do* exist—but *as* complexes of referents of the more basic theory *A*. Having paid our respect to the theoretical attitude and its findings, we continue operating in accord with the practical attitude, without fear of having *nothing* to talk *about*. We may, therefore, affirm without contradiction all of the following:

(1) *B* is reduced to *A*. (The accomplished reduction.)

(2) Only the referents of *A* exist at bottom. We have indeed *replaced* one domain by another. The ontology of *A* is fully adequate for describing those phenomena previously accounted for by the ontology of *B*. (The theoretical attitude.)

(3) We may still speak (correctly) about the referents of *B*. (We are not talking nonsense when adopting the practical attitude.)

(4) There is but one world—one set of phenomena—and two ways of describing it. (We keep both theories for the domain in question.)

From the theoretical point of view, in which alone ontology is at issue, the reduced theory B and its basic concepts are *replaced* by A; the apparatus of B is kept "only" for what I have called practical purposes.

Do electric fields exist or don't they? Yes or No? Bad question. The proper response is: From a *theoretical* point of view, electric fields are (at best) superfluous; we can fully describe any phenomena in the old domain of electrodynamics—any phenomena previously accounted for with electrodynamic laws—on the basis of the apparatus (including the ontology) of the remainder of mechanics. Electric fields may be replaced by a complex mechanical structure *for this purpose*. But it may still be more useful or appropriate for certain *other* purposes to continue to use the electric field concept in its original domain. What (if anything) are we talking about when we do so? At bottom, only complexes of mechanical entities. Do the alleged electric fields exist? As mechanical structures, Yes; but as a kind of entity distinct from and correlated with the latter via "bridge laws," No. What is intertheoretical reduction? From the theoretical point of view, it *is* replacement; from the practical point of view, it *may* be construed as involving identification.

Theoretically, we "keep" only theory A and its vocabulary and interpret the replacement proposals as metatheoretic in the manner suggested. But in contexts appropriate to the practical point of view, we recognize the great convenience of keeping and using both theories. *In these situations*, we may if we wish read the replacement proposals as proposed identifications. This allows one, in the object language, to state a set of bridge "laws" in the form of identities. But it is only in these practical contexts that "bridge laws" can be formulated in the object language. One must not forget the problems that result from misconstruing this identity interpretation; the practical attitude must, on reflection, give way to the theoretical—in which reduction is replacement.

One Further Example

There are two more things I wish to do before I go on, in the next chapter, to consider a type of intertheoretic reduction for which the concept replacement analysis appears to be inappropriate.

First, I wish to clarify further and to reinforce some of the points I have been making by sketching how my analysis applies to one more

example—that of the reduction of classical *thermodynamics* to statistical *mechanics*.   This case, unlike the electrodynamics/mechanics
example upon which I have so far leaned, has the (not unalloyed)
advantage of being a *successful* reduction.   Secondly, I wish to consider
the implications which my account of reduction has for our understanding of the related notion of scientific *explanation*.

The thermodynamics/statistical mechanics case has been a standard
example in the literature on reduction.   I have referred to it briefly, in
Chapter 3.   It is plausibly interpreted as successfully providing replacement functions for the basic terms of thermodynamics in terms of
complex expressions drawn from classical dynamics.[8]   The usual
example is the concept of *temperature*, which, for certain types of
gases, is replaced by a function of the *masses* and *velocities* of the
molecules constituting the gas (i.e., discrete particles or complexes
thereof, interacting in accordance with various force laws).

It would, however, be more accurate to say that classical thermodynamics is reducible to classical dynamics plus a *matter theory*, i.e.,
statements regarding the existence of objects of a certain nature (mass,
size, etc.)—molecules—conforming to the general laws of dynamics *plus*
statements of the force laws in accordance with which these objects
interact.   These additions are absolutely essential, for the general laws
of motion of classical dynamics carry by themselves no implication
regarding the *existence* of anything conforming to them.   As far as the
laws of classical dynamics are concerned, Descartes might well have
been right in postulating a plenum as the basic "matter theory," in
which case applied fluid mechanics would be the correct detailed
matter theory.   This neglected but crucially important point is an
example of what I had in mind when I spoke, in Chapter 3, of a scientific theory containing a view about the kinds of *entities* constituting
and functioning in a domain.   So, classical thermodynamics is reducible
to classical dynamics plus a particular matter theory—namely, "atom
theory" (as opposed to a Cartesian plenum).[9]   The reduction is analyzable as a successful replacement of each thermodynamic term by
a complex of terms from classical dynamics plus what I have called
"atom theory."

I have been considering gases only, but classical thermodynamics
applies to other states of matter as well.   More generally, thermo-

dynamics can be described as the theory which deals with the *molar* thermal (and related) properties of matter without attempting to account for them on the basis of the *micro*structure of the systems under consideration. (It has been called a "phenomenological theory" for this reason.) In this more general sense, thermodynamics is reducible not to the classical statistical mechanics of gases (the "kinetic theory of gases") but to contemporary theories about the structure of matter in general—for instance, quantum mechanics. Thus understood, the complexities involved in the reduction are enormous; one appreciates, here, why the deducibility of laws cannot be a necessary condition of reducibility. I have already quoted from a physical scientist in this regard, in Chapter 3; here is the view of another, who writes about the relation between macro- and micro-theory for systems other than gases:

> In the present state of development general equations exist from which one can *in principle* [my emphasis] compute equilibrium thermodynamic properties from molecular models. These equations have been solved for many systems of independent or almost independent particles, such as perfect gases. . . . The difficulties of the equilibrium theory lie in the mathematical problems encountered in almost every investigation of systems of interacting particles— liquids, . . . etc. Mathematicians have not developed the required theory of functions of many dependent variables. . . . It is not clear how to formulate many questions which can be asked about systems far from equilibrium. . . . A completely rigorous development [of equilibrium statistical mechanics] is quite lengthy and in many ways not yet available.[10]

There is a further point which the present example helps to reinforce. The laws which flow from the basic dynamical laws are, when applied to a gas, *statistical* in nature, because the information supplied regarding the distribution of molecules, their velocities, etc. is statistical. But the laws of thermodynamics are not. Hence, it might be said that, in such a case, the laws of thermodynamics are *not* strictly deducible from, *or transformable into*, the laws of the reducing theory. Should we say that the reduction does not succeed, after all? No, for we can say that the laws of the reduced theory are *not quite correct* in the relevant regard (although they are certainly well within experimental error for most cases in which they are actually applied). There are good reasons for proceeding thus. The number of molecules in ordinary-sized samples of gases and other materials is so large that statistical irregularities introduced at the basic level of the motions of molecules in the reducing theory are obscured beyond detection at the level of the reduced theory. In Chapter 1, I discussed the problem of what is to be done with those parts of theory B that seem not to accord with the reductive program (analyzed as concept replacement, so that the laws of B transform into laws of A). Here, we have a specific case conforming to my earlier observations on this issue. With good reason, we say that theory B (thermodynamics, in this case) is simply *wrong* in the respect in question: the attempted reduction of B to A makes this clear. Here, then, is a concrete example of the sense in which a "totally successful" replacement is but an ideal that we know we cannot achieve.

We may describe the ontological implications of this concept replacement reduction as follows:

(1) Classical thermodynamics (of gases) is reduced to classical dynamics plus "atom theory." The basic concepts of the former are replaceable by those of the latter (or by some more current theory, such as quantum mechanics).

(2) Only the referents of classical dynamics plus atom theory exist at bottom, in the domain of thermodynamics, i.e., atoms and molecules interacting in a manner analyzable in accord with the laws of classical dynamics.

From the theoretical point of view, there is no further need for the ontology of thermodynamics. Classical dynamics plus a "matter theory"—hypotheses about the constitution of gases—provide a fully adequate ontology for describing what is taking place in the original domain of classical thermodynamics. For example, we may now speak of the *mean kinetic energy of the molecules* instead of the *temperature* of a sample of gas. This may be put more forcefully by saying that temperature doesn't exist, *meaning thereby* that classical dynamics plus "atom theory" has been found to be fully adequate for the description of those phenomena for which the concept of temperature had previously been used—the old domain of thermodynamics. Additional problems probably concern (i) the particular molecular constitution of particular kinds of gases, and (ii) mathematical complications regarding the detailed deductions discussed earlier.

Nevertheless,

(3) We may still speak correctly about the referents of the terms of thermodynamics for certain "practical" purposes. Temperature still exists, if you like (insofar as we wish to speak of the *existence* of properties or quantifiable characteristics), but it exists only as a complex of referents like mass and velocity.

(4) There is but one set of phenomena involved here—the behavior of gases—and two descriptions of it, molar and micro, one of which is fully adequate for all theoretical purposes, the other of which is convenient for other purposes.

It might be argued that it is only by historical accident that we have a science of thermodynamics. Had an acceptable well-articulated "atom theory" of quantitative form been introduced first, thermodynamic concepts might never have been introduced at all—or introduced specifically as shorthand expressions for complexes of terms from classical dynamics plus atom theory. One can, in fact, maintain this position despite a related claim that seems to contradict it. Robert Rosen, whom I quoted in Chapter 3, maintains for instance that

> . . . if the gas laws had not been known *first*, they would never have been discovered through statistical mechanics alone. Formalism will indeed enable you to form any averages you want, but it will not tell you what these averages mean, and which of them are useful and important in specifying and describing macrosystem behavior. (*op. cit.*, p. 187)

Howard Pattee, in a paper (appearing in the same volume) entitled "Primitive Functional Hierarchies," makes essentially the same point when he speaks (on p. 169) of

> . . . the inability of the formal mathematics to predict what collective properties of complicated systems will produce simple, significant effects in the physical world of the observer. In other words, while there is no question that the detailed equations of dynamics can be used to calculate previously well-defined averages or collective properties, there is no way to predict from only the dynamical laws of the system which definitions of collective properties are significant in terms of what we actually can measure.

One can admit the force of these remarks while at the same time conceding the superfluity of thermodynamics from the *theoretical*

point of view. This is because the claim that a theory is fully adequate
for the description of what is taking place at bottom in a domain carries
no implications regarding the "significance" to observers of particular
applications or other related consequences of that theory. This is not
to denigrate the practical attitude, only to recognize that it differs from
the theoretical attitude.

It may even be the case that the theoretical attitude requires information generated with the aid of the practical, in order to achieve some
of its best results. After all, we test our theories by deducing predictions from their laws. It may well be, as Rosen and Pattee maintain,
that we must *already have* the gas laws before realizing that we *ought*
to deduce them from the basic dynamical laws. Instructive though this
may be, it does not contradict the fact that the gas laws have actually
been deduced.

Finally, if the point that Rosen and Pattee advance is valid at all, it
is valid *within* classical mechanics itself (indeed, within *any* theory).
There may be many laws from different branches of mechanics which
must be "known first," in the sense that they might otherwise never
have been discovered by means of the basic laws of mechanics alone.
Perhaps Kepler's law of planetary motion to the effect that equal areas
are swept out in equal times is such a law. How else might physicists
have fist realized that the area swept out is a "significant" feature of
planetary motion worthy of further study (as opposed, say, to the cube
root of the brightness of the planet)?   Their point, then, however
important, is much too general to enable one to draw specific conclusions about particular theories or theory-pairs. It comes simply to this:
given the basic laws of a theory and *no other information whatsoever*,
you can have no idea of what to do next! No amount of mathematical
skill will tell you which deductive consequences or lines of thought are
or are not *worth following* among the infinite possibilities. Thermodynamics is, in this regard, no worse off with respect to classical
mechanics (despite the occurrence of averages and "collective properties") than any consequence or branch of any theory with respect to
the fundamental hypotheses of that theory.

If this finding is correct, my ontological claim survives with even
greater plausibility: *When one theory has been reduced to another,
one need no longer assume the separate existence of entities referred
to by the basic terms of the reduced theory.*   From the theoretical

point of view, the ontology of the reducing theory is fully adequate in the original domain of the reduced theory. (The reduced theory may be viewed simply as a branch of the reducing theory. See Chapters 7, 8.)

William Wimsatt has, with respect to the mind-body problem, made a point similar to Rosen and Pattee's. He considers elimination (replacement) of the mental realm in favor of neurophysiology to be

> . . . unlikely—indeed virtually impossible—because neurophysiology cannot make progress at the level of higher units of functional organization without appealing to the mental realm *for guidance*. The task would be like asking a molecular biologist to give a molecular reconstruction of elephant physiology from what he knows, together with photographs of elephants taken in the wild at a conservatively safe distance. Constant reference to minute details of the upper level descriptions of elephant anatomy and physiology are at least necessary (though not yet sufficient) for the task. The rebirth of cognitive psychology . . . [is] absolutely essential for neurophysiological progress. We must *use* our current and future intuitions and theories of the mental realm to *help to develop* our neurophysiological accounts. [emphases mine] [11]

Here too, I think, it could be argued that the *need* for the upper level theory is what I have called a *practical* one rather than a theoretical one—not a matter of fundamental ontology. But the kinds of reductive issues which arise in the mind-body problem are perhaps sufficiently different from those within physical science to prevent any facile application of my analysis of reduction—especially my ontological claims.

Reduction and Explanation

My analysis of reduction has implications for our understanding of the concept of *explanation* in science. I have already touched on these, but more may usefully be said about the relation between reduction and explanation. According to the standard analysis, all theory *reductions are explanations*. But if theories are viewed as conjunctions of laws, and if, in turn, the explanations of laws are viewed as the deductions of laws, then all reductions of theories require deductions of laws.

In brief, *reduction requires deduction*. According to the replacement analysis, however, some reductions (of theories) do *not require* deductions (of laws). Hence, since I maintain that reduction (of the kind thus far considered) constitutes a type of explanation, it becomes an important consequence of the replacement analysis that *explanation does not always require deduction*. In effect, the replacement analysis of reduction conflicts with the deductive-nomological analysis of explanation. (This conclusion could be avoided by maintaining that reductions are not necessarily explanations, but I think that, too, would be incorrect.)

It will be useful, here, to draw some distinctions not usually made: one between the reduction of *theories* and the reduction of *laws*; another between the *reduction* and the *deduction* of laws. In order to have an element of concreteness in our account, let us suppose that we have before us two theories A and B such that B has been reduced to A, in the sense of the concept replacement analysis. Let L_b be a law of theory B; and let L_a be the statement in the language of theory A into which L_b transforms when all B-terms in L_b are replaced by complexes of A-terms, in accordance with the appropriate replacement functions. I am assuming that theory B has been reduced to theory A, which means that the statement L_a is a *law* of A. Recall that this does *not imply* that L_a has been deduced from the basic laws or postulates of theory A. Assume, in fact, that L_a has not been thus deduced but has been verified by some other means described earlier on.[12]

We may now ask two distinct but closely related questions:
(a) Has the *law* L_b been *re*duced to the laws of theory A?
(b) Has L_b been *de*duced from the laws of theory A?

The answer to (a) is clearly Yes. For, with the aid of the replacement functions, L_b transforms into a statement L_a, which can be shown to be a law of theory A (or at least a verified statement of theory A). Let us linger over this a moment. The law L_b describes an aspect of the behavior of entitites referred to with the aid of terms from the vocabulary V_b of theory B. Reducibility of *theory* B to theory A means that the phenomena in question (and others) can be successfully *redescribed* as the behavior of (more complex) items *in the ontology of theory* A. (What we previously referred to as B-items are really at bottom complexes of A-items; or, where we had believed the former to obtain, we now know the latter does.)   From the theoretical point of view, one need no longer use L_b to describe the phenomena in question; L_a will serve in its stead.   I do not see what more could reasonably be required in order to say that the *law* L_b has been reduced to the *law* (or verified statement) L_a.   If one were, in spite of this, to maintain that the law L_b has *not* been reduced to the law L_a (because of the non-deducibility of either from the postulates of theory A—or for any other reason), the proper response would simply be that *theory* reduction does not require *law* reduction!   That is, inhomogeneous theory reduction *is* successful entity or concept replacement, as certified by establishing in *some* way or other that the laws of B transform into laws (or verified statements) of A, whether or not one refers to the non-deductive certification or establishment of the relevant *laws* as *their* reduction.

What about question (b)?   Has L_b been deduced from the laws of theory A?   In a restricted sense, the answer might be Yes.   For, if we use each replacement function to justify writing down a certain related biconditional (or identity) statement in the object language common to theories A and B—"bridge laws," then L_b can be deduced from the conjunction of L_a with these bridge laws.   (This is just part of the description of reduction according to the standard analysis.)   I shall not argue whether or not it really is proper to refer to this relation between L_a and L_b as the *deducibility* of the latter from the former.   In a more interesting sense, the answer to question (b) is No; for, by hypothesis, L_a has not been deduced, in turn, from other laws of A—in particular, from the *postulates* of A.   (Those who adopt an account of laws in which deducibility is essential would *have to* conclude that L_b has *definitely not* been deduced from the *laws* of theory A.)   Hence, law L_b may be said to have been *re*duced to the laws (or verified statements) of theory A—in particular to L_a—even though it may be false or at least misleading to say that L_b has been *de*duced from the laws of theory A.

Now, a third question arises:

(c)   Has the law L_b been *explained* on the basis of theory A, in virtue of having been *re*duced without having been *de*duced?
I think the answer is clearly Yes.   The phenomena previously described by L_b have been accounted for by showing how they can be correctly redescribed on the basis of the conceptual apparatus of a deeper theory, A.   If this isn't a brand of explanation, then I do not

know what is. But the point at stake is more fundamental than that of the deduction of L_b. It hardly matters whether one wishes to say that L_b has or has not been "deduced" from the "laws" of theory A. For whether it is correct or not, the crux of the issue of *explanation* does not lie with that. In a *reductive explanation* of the kind under consideration, it is, rather, our certified ability to *redescribe on the basis of concepts of a deeper theory* which is the heart of our gain in understanding. And, here, it does not matter where the certification of the correctness of the redescription comes from.

If a physical theory is viewed as an *explanatory scheme*, then the present point may be expressed by saying that, in inhomogeneous theoretic reduction, one explanatory scheme is replaced by a deeper one. The explanatory function of concept replacement reductions is different from, and deeper than, that of explanation *within* any given explanatory scheme or theory (in which it may well be the case that deduction *is* required—a matter I shall not discuss here).

This important difference between explanation within a theory and explanation (reduction) *of* an entire theory shows itself as well in the following interesting way: L_b has been reduced to, and thus explained on the basis of, L_a, even though, because of its non-deducibility from the postulates of A, L_a itself may be wholly unexplained (as yet) *within* theory A. We may for example explain what was previously considered an electromagnetic phenomenon (assuming the reductive program to have been successful) by showing how it really is a type of rather complex mechanical phenomenon involving (say) elasticity, even though phenomena of the latter type have not been fully explained on the basis of—deduced from—the basic laws of the remainder of mechanics.[13] This demonstrates quite forcefully how deducibility *may* function as a necessary condition of explanation *within* a theory (*if* the deductive-nomological analysis of explanation is basically correct) though it be at best only a sufficient condition for testing whether a *theoretic reductive explanation* has been achieved. Nagel is correct in maintaining that inhomogeneous theoretic reductions are explanations, but they are not entirely like explanations within a theory.

The deduction of laws and the related endeavor to achieve deductive unity for each of our theories are, admittedly, significant features of the scientific enterprise (especially of mathematicised sciences like physics). But acknowledging that does not preclude the relevance and importance—the at least equal importance—of other features of science. Our own topic is one of these, of course, the perennial attempt to replace one way of describing a given domain—one "explanatory scheme"—by a deeper and more general one. This constitutes another kind of striving toward unity—*ontological unity* rather than deductive unity. It is part of the attempt to use the smallest number of *concepts* to organize our knowledge of the world; the attempt to achieve deductive unity is part of the attempt to use the smallest number of *laws*. The standard analysis of reduction has, in my view, overemphasized the importance of the latter. I have, on the contrary, attempted to reassert the importance of the former.

Notes

1. See his "The 'mental' and the 'physical'," in H. Feigl, M. Scriven, and G. Maxwell (eds.), *Minnesota Studies in the Philosophy of Science*, Volume II, Minneapolis, 1958, pp. 382, 428.

2. Just as the relation 'has more terms than' holding between two theories does not generate any law requiring explanation *within either* of the theories so related (or any third theory).

3. See Robert Causey, "Uniform Microreductions," *Synthese*, Vol. 25, nos. 1/2, 1972, pp. 176-218.

4. See Tryg A. Ager, Jerrold L. Aronson, and Robert Weingard, "Are Bridge Laws Really Necessary?," *Nous*, Vol. 8, No. 2, May, 1974, pp. 119-134 (especially pp. 129-130). For Causey's response, see "Identities and Reduction: A Reply," *Nous*, Vol. 10, No. 3, Sept. 1976, pp. 333-337 (especially p. 337, note 1).

5. Statements like this tend to indicate that Causey's solution to the explanation problem could not work in all cases. According to Causey, bridge "laws" do not need to be explained because they are really *identities*. But there is no plausible way to turn this statement into an identity. The left side refers to the *presence* of a thing; the right side refers to the *behavior* of various things. No attributes, or things, are being identified. (For more on this point, see Judith Jarvis Thomson's "The Identity Thesis," in the Nagel Festschrift referred to earlier. My analysis circumvents this kind of criticism.

6. Perhaps an overemphasis on the practical attitude is one important element involved in prominent physicalist and phenomenalist (generally, instrumentalist) interpretations of physical theories. It is often via successful predictions of the practical sort that one establishes

the correctness of the physical ontology claimed in accord with the theoretical attitude. Overemphasizing such predictions leads to such excesses as the verification theory of meaning. Perhaps I should say, rather, that the two attitudes *need not* be mutually exclusive.

Incidentally, one should not equate 'practical' with 'pragmatic'. What I earlier referred to as *pragmatic* features of a theory includes both the theoretical and practical attitudes: the attitude that a scientist has toward a theory is part of the "pragmatics" of that theory.

7. Compare my earlier remarks about rational and irrational number theory in Chapter 1.

8. See Chapter 3.

9. Do not confuse this coined term, 'atom theory', with the expression 'atom*ic* theory', which has a different usage in current physical science. For more on this point, see Chapter 5. (Current atom*ic* theory is one kind of "atom theory." Democritus had another kind.)

10. E. W. Montrall, "Principles of Statistical Mechanics and Kinetic Theory of Gases," in E. U. Condon and H. Odishaw (eds.), *Handbook of Physics*, New York, 1958, Part 5, p. 11.

11. See his very interesting paper "Reductionism, Levels of Organization, and the Mind-Body Problem," in G. Globus, G. Maxwell, and I. Savodnik (eds.), *Consciousness and the Brain: A Scientific and Philosophical Inquiry*, New York, 1976, p. 236.

12. Incidentally, some would argue that for it to be a *law* rather than a "mere generalization," a statement *must be deducible* within a theoretical system. (See, e.g., R. B. Braithwaite's *Scientific Explanation*, Cambridge, 1953, Chapter 9.) Those who sympathize with this analysis of the concept of *law* may read 'verified statement' wherever I use 'law'; it is really only the notion of a *verified statement* that is essential to my analysis of theory reduction. Verified statements of theory *B* must transform into verified statements of theory *A* (whether or not they are considered to be "laws").

13. There may even be cases of reduction without a deeper theory. These arise in situations in which it is best to speak directly of *entity* reduction (rather than of theory reduction). I will have more to say about this, in Chapter 6.

Chapter 5

Direct Theory Replacement

I have now completed the direct examination of one kind of intertheoretic reduction, "concept replacement reduction." Later, I will show how the analysis provided can be used to illuminate certain related reductive notions, in particular, how one entire *branch* of science may be reduced to another. Before doing so, however, I should like to consider another type of *theory* reduction found in physical science—one which differs in certain essential respects from the kind of reduction I have been discussing in the last four chapters. It will appear that the concept replacement analysis does not, as it now stands, provide an adequate analysis of the kind of reduction I now have in mind. As we shall see, the "standard analysis" may appear to fare better, although it is just in reductions of this new type that it has been most heavily criticized in the current literature. The key idea in this type of reduction is still *replacement*, but it is direct *theory* replacement rather than theory replacement via an item-by-item *concept* replacement. Here, we shall be dealing primarily with theories *competing for the same* domain rather than with the notion of one domain *underlying*—and thus replacing—another domain.

We may begin by noting that the standard analysis of reduction has usually been attacked on grounds other than the ones I have already developed. One such criticism holds that, in certain paradigm cases of reduction, the laws of the reduced theory B are *not deducible* from those of the reducing theory A; rather, one can only deduce from A *approximations* of the laws of B in a clearly specifiable sense. This is a clearly different type of non-deducibility than the kind I discussed in the last two chapters; there is no problem of mathematical or experimental complexity in the type of case I now have in mind (or when there is, it is superimposed on the present characteristics).

Hempel has considered such an objection to the deducibility condition; he appears to have disarmed it simply by accepting the objection while denying its importance. He holds that no more would be needed here for a reasonable notion of reducibility than that ". . . the new [this term will be important later] or reducing theory . . . be required to imply that, within a certain domain, close approximations of the old theory hold good."[1]

Critics of the standard analysis seem to have failed to notice that, as early as 1956, in Kemeny and Oppenheim's pioneering work on reduction, this objection had already been considered and disposed of similarly. They wrote then as follows:

> . . . the old [this term will be important later] theory usually holds only within certain limits, and even then only approximately. For example, in the reduction of Kepler's laws to Newton's we must restrict ourselves to the case of a large central mass with sufficiently small masses, sufficiently far apart, around it. And even then the laws hold only approximately—as far as we can neglect the interaction of the planets. While these points are of fundamental importance, there is no way of taking them into account as long as we tacitly assume that our theories are correct. If we abandon this (contrary-to-fact) assumption, then the problem of reduction becomes hopelessly complex. (*op. cit.*)

These authors simply accept, therefore, the deducibility condition, but they explicitly construe it as an "oversimplification" (*ibid.*). Incidentally, I should mention that they also thank Hempel for "clarifying their thinking on this point" (*ibid.*).[2]

There is a certain irony in the way in which the issue arises in the present kind of case as opposed to those discussed in the previous two chapters: for, once their approximative character is conceded (as Hempel concedes, above), the deductions involved are extremely *easy* to carry out. On the other hand, in concept replacement reduction, the deductions involved may be extremely *difficult* to carry out, even when approximations are allowed. It will be useful to display some examples of this type of reduction before considering it in a general way.

I shall present only as much detail as is necessary to show how different in character it is from the kind of reduction previously discussed.

Some Examples

(1) Special Relativistic Mechanics / Classical Newtonian Mechanics

The relation between relativistic and classical mechanics presents a clear instance of the type of reductive situation at issue here. The laws of classical mechanics are not actually deducible from those of relativistic mechanics; instead, one can derive only *approximations* of the laws of classical mechanics as certain parameters of relativistic mechanics are allowed to approach certain limiting values. Roughly, if one considers systems in which velocities are small relative to the velocity of light, the relativistic laws describing such systems approximate the classical laws for such systems:[3] as $v/c \to 0$, relativistic laws $\to$ classical laws. The "deduction" of the laws of the reduced theory from the laws of the reducing theory is, then, of a rather special type. Moreover, deductions of this sort are very easy to carry out, as I have been claiming. As one lets v/c approach zero in the relativistic laws, the corresponding classical laws quickly emerge, with very little mathematical manipulation. (See, for example, the force laws for the two theories, below; even the mathematically unsophisticated reader can carry out the derivation.)

Thomas Nickles has pointed out that, in such cases, the relation of reduction, using the term 'reduce' as it is actually used *by physicists*, goes in the reverse direction: as $v/c \to 0$, relativistic mechanics is said to "reduce to" classical mechanics rather than the other way around![4] As we shall later see, this constitutes what Nagel has called a "homogeneous" reduction. Classical mechanics contains no terms which do not also appear in relativistic mechanics, so no "bridge laws" are required. In the relevant contexts, then, no extra terms appear in theory B that need to be replaced by (or identified with) terms selected from theory A. (Of course, the notion of concept replacement may be applied in a trivial way: we may "replace" each term of the reduced theory *with itself*.)

(2) Quantum Mechanics / Classical Mechanics

It is difficult to say something at once interesting and non-controversial about the relation between quantum mechanics and classical mechanics. But it may plausibly be said that that relation is quite similar to the one here being considered. The laws of classical mechanics are not deducible from the laws of quantum mechanics; but, once again, approximations of the classical laws may be derived if certain parameters of the quantum mechanical laws are allowed to approach certain values. Thus, if certain specified parameters are allowed to become large relative to Planck's constant h (that is, if we consider systems with large quantum numbers), then the quantum mechanical laws approach the classical mechanical laws for the relevant systems. Each law of classical mechanics can then be obtained by choosing appropriate limits with respect to the appropriate quantum mechanical laws. (Once again, the derivation is mathematically easy.) In fact, the relationship noted suggests an interpretation of the so-called "correspondence principle." As one author remarks:

> . . . the principle is that for large quantum numbers—large values of energy and momentum or of "action" as measured in units of h—the quantum theory will approach classical theory as an asymptotic limit.[5]

This is to say, in effect, in the spirit of Nickels' observation, that, under certain limiting assumptions, quantum mechanics reduces to classical mechanics—rather than the other way around.

It appears then that the much-discussed "correspondence principle" may simply be viewed as a reducibility claim for a specific pair of theories. If so, then similar "correspondence principles" may be generated for pairs of theories wherever one is reducible to the other in the manner being considered. Indeed, Mario Bunge interprets the correspondence principle in just this way; he writes:

> Correspondence principles—in [statistical mechanics, special relativity, general relativity, quantum mechanics and quantum field theory] —are conceptual tests for the compatibility of those theories with less refined theories. . . . The correspondence principle for [quantum mechanics] . . . is N. Bohr's unperishable contribution to metascience. . . .[6]

I shall not pursue this possibility further, for we need not entangle ourselves here in the thicket of the philosophy of quantum mechanics.

Such a reduction, however, like the one previously considered, is an instance of a homogeneous reduction. Classical mechanics contains no terms that are not also contained in quantum mechanics. Once again, therefore, no "bridge laws" are required. But in this case, there are terms that occur in the reducing theory (e.g., the wave function Ψ) that do not occur in the reduced theory (that is, classical mechanics). This was not true before: special relativistic mechanics contains no terms that do not also occur in classical mechanics. It should be borne in mind that the distinction between homogeneous and inhomogeneous reductions only concerns the question whether or not the reduced theory contains terms contained in the reducing theory. No one, to my knowledge, has yet investigated whether the converse distinction is significant. It seems intuitively clear that reductions in which the reducing theory contains terms not occurring in the reduced theory are significantly different from reductions in which this is not the case. But I will not pursue the issue here. At any rate, there may be important differences *among* reductions of the present type. Other differences among the examples given will emerge as we proceed. It should be clear, therefore, that my division of reductions into two distinct types is not meant to mask interesting and important differences among reductions within each type.

We now have before us a situation interesting enough to merit closer attention. We have a theory, classical mechanics, which can be reduced independently to two distinct theories, relativistic mechanics and quantum mechanics. The situation may be diagramed in an obvious way:

arrow indicates reduction
in *Nickles'* sense

No particular difficulties arise here, since both A and A' (relativistic and quantum mechanics) deal with those aspects of nature with which B (classical mechanics) is concerned. Each, however, is "refined" (Bunge's term) *in its own distinct way*; each *corrects B*, but in different respects.

We can also combine the refinements of A and A' into one comprehensive theory, viz., *relativistic quantum mechanics*. That theory reduces (in Nickles' sense) to quantum mechanics (or to "ordinary" or "non-relativistic" quantum mechanics), as $v/c \to 0$. It may seem some-what odd to speak similarly of reducing (Nickles) relativistic quantum mechanics to special relativistic mechanics. It seems so because, roughly, relativistic quantum mechanics is not merely a conjunction of A and A'. The designations 'relativistic' and 'quantum' function at different methodological levels: the adjective 'relativistic' designates certain space-time transformation properties of the equations of quantum mechanics (or of any other theory to which the term 'relativistic' is similarly applied). If by the expression 'special relativistic mechanics' we mean *that theory of motion and of its causes*—that *dynamical* theory—which replaces Newton's and conforms to the special *principle* of relativity and which is therefore Lorentz-invariant,[7] then we can diagram the reductive situation as follows:

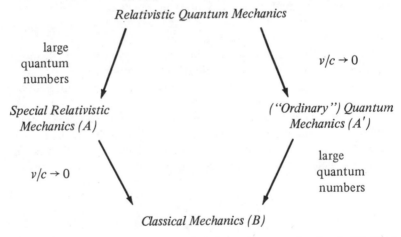

Relativistic Quantum Mechanics

large quantum numbers

$v/c \to 0$

Special Relativistic Mechanics (A)

("Ordinary") Quantum Mechanics (A')

$v/c \to 0$

large quantum numbers

Classical Mechanics (B)

(As in the previous diagram, the arrows represent reduction in Nickles' sense.) The diagram clearly exhibits what is evident in its own right: the reduction relation under consideration is *transitive*. It also shows something not so obvious: the reduction relation is not *connected*; A does not reduce to A' and A' does not reduce to A.[8]

(3) General Relativity / Special Relativistic Mechanics
(plus classical gravitational theory)

General relativity theory reduces (in Nickles' sense) to special rela-tivistic mechanics plus classical gravitational theory under certain limiting conditions: little or no acceleration, "small" regions of space-time.   Or, following the terminology of the standard analysis of reduction, special relativistic mechanics and classical gravitational theory each reduce to general relativity theory; *approximations* of the former two can be deduced from the latter when certain parameters are allowed to approach certain limiting values. So this example resembles in relevant respects the two examples just considered.

But there are differences. Unlike the previous two examples, this one concerns an inhomogeneous reduction.   The reduced theories, special relativistic mechanics and classical gravitation theory, each contain a key concept, *force*, which does not occur in general relativity—that is, in the reducing theory. In this respect, the example shares an important feature with the examples of concept replacement reduction discussed in the first four chapters. Hence, my distinction between the two types of reduction is somewhat idealized in the sense that there may be cases of reduction sharing features of both types. I shall say more about this, below. In the present case, the reducing theory contains no concepts not contained in (or at least not easily expressible in) the reduced theory; General Relativity introduces no new entities (although of course the old entities are treated in rather different ways).

Direct Theory Replacement Reduction

It is time to discuss the type of reduction illustrated by the foregoing three examples in a more general way.

All three have the following features in common (unlike those discussed in the first four chapters, which have none of these features):

(a) The notion of an *approximation* is involved in an essential way. It is *limit-taking* in a precise sense sufficient for getting from the reducing theory to the reduced theory in each case.

(b) The deducibility condition of the standard analysis of reduction requires modification to handle these cases—perhaps a minor modification (one which Hempel accepts), but a modification nonetheless.

(c) A description using 'reduces to' in the reverse sense is equally appropriate and is favored by physicists.

Let me emphasize that none of the reductions discussed in the earlier chapters had any of these features. So these form a truly distinct species of reduction (notwithstanding other significant differences that may be found among members of the species). To show this more clearly, I shall try to answer the question: How does the *replacement analysis* of reduction fare in these cases? Recall that that analysis involves *theory* replacement *via concept replacement* in a *term-by-term* manner. Let us proceed case by case, beginning with the special relativistic mechanics/classical mechanics example. Here, there seems to be *no* replacement of concepts or terms (or entities) at all. This is a homogeneous reduction. Now, there are authors holding certain views regarding *radical meaning variance* who would claim that all the relevant terms *do* change in meaning—that we *do* have a *new set of concepts* in relativistic mechanics. I have argued elsewhere[9] that, in this case, their claim is simply not true. I have shown, for example, that it is much more plausible to maintain that the term 'energy' has the *same* meaning in special relativistic mechanics and in classical mechanics. The arguments need not be repeated here. (More will be said in the next chapter about the use of "theories of meaning" in the philosophy of science.) But even if it were true that terms like 'energy', 'mass', etc. have different meanings in relativistic mechanics and in classical mechanics, it remains true that all the *terms* of these two theories are the same (whether or not they have "different meanings" and hence whether or not they express different *concepts*). There is, moreover, no replacement, here, of terms (or concepts) of B with *complexes* of terms (or concepts) of A. These considerations are sufficient to confirm the difference between the two types of reduction I am insisting upon. The reductive situation here is best described as one in which the laws of theory A involve a partial *rearrangement* of terms in the laws of theory B such that the arrangement in B may be mathematically recovered from the arrangement in A when certain limits are taken. (See Chapter 10 of my book on relativity theory for further details on this point.)

So we do not have a case of reduction here that may be analyzed fruitfully as a *term* or *concept* replacement reduction. But the notion of *replacement* is still central. This sort of reduction may be viewed— and is explicitly so viewed among physicists—as a successful attempt at a *theory replacement*, a *direct* theory replacement rather than one resulting from an item-by-item term or concept replacement. Historically, special relativistic mechanics was introduced *as a whole* to *replace* the current *theory* of motion. There was no item-by-item research, no extended period during which, concept by concept, the older theory gave way to the newer.

How would the *concept* replacement analysis work for the quantum mechanics/classical mechanics case? It seems to work to some extent. For example, the concept of a particle—a localizable mass having a specific and determinable trajectory—is *replaced* by a complex function of terms involving the concept of a wave function Ψ. (This is not uncontroversial, however.[10]) Nevertheless, it is only in conjunction with the "correspondence principle," which involves *taking limits* in a certain way, that "the laws of B transform into the laws of A" upon the employment of the appropriate concept "replacement functions." There are also elements of *term rearrangement* involved. Still, the most instructive way of viewing this reduction is, once again, as a *direct theory replacement*. The early history of quantum mechanics involves a succession of attempted replacement-*theories*, whereas the electrodynamics/mechanics case is better described as a succession of attempted *concept* replacements (*resulting* in associated *theory* replacements). Various versions of quantum mechanics were originally introduced and defended primarily as replacements for the current *theory* (classical mechanics including electrodynamics), which ran into serious difficulties in attempted applications to atomic and sub-atomic systems and phenomena.

Finally, what about the general relativity/special relativistic mechanics plus classical gravitational theory case? Here, too, a direct *theory* replacement obtains—one theory of motion and gravitation replaces an older theory covering the same domain. But there are also elements of concept replacement here. The concept of *force* in the older reduced theories is replaced by a complex function of parameters from the new theory (involving space-time structure). In this case, however, the space-time parameters occur in the older theories as well; they are just differently arranged there. Forces are *replaced* by a function of space-time curvature; and from the perspective of General Relativity Theory, may be said not to exist at bottom—which recalls the "theoretical attitude" of which I spoke, in Chapter 4. The partial resemblance of this case to concept replacement reductions is, therefore, a resemblance with an interesting twist, for it could be said that general relativity shows that *force* in classical mechanics can be replaced by a suitable combination of other classical mechanical concepts. The replacement is not effected simply by selecting entirely different concepts from a new theory covering an "underlying" domain as it is, for instance, in the case of thermodynamic concepts vis-à-vis the kinetic theory of gases.

In each of the examples of reduction just considered, as in those developed in earlier chapters, *replacement* is the central idea. Here, it is a matter of *direct theory replacement*, although concept replacement occurs to some extent. A good part of the motivation is this: in order to handle the *same domain better*, the original theory, *seen to be in trouble*, is replaced in its entirety by a *new* theory involving for the most part the *same* concepts.

This is worth emphasizing. The reduced theory is found or believed to be *incorrect* within its domain: it is *for that reason* to be *replaced* by another theory thought to be correct—correct at least in those respects in which the reduced theory failed—*without probing deeper, necessarily, to an underlying domain* (but also without precluding such an effort).

This contrasts with the concept replacement reductions already considered. It was *not* because of difficulties in thermodynamics or electrodynamics that physicists sought to reduce each to *deeper* level theories. In those cases and in others where concept replacement analysis seemed more appropriate, the reducing theory was not originally *introduced* to perform a reductive role; the theories involved existed side by side (indeed, sometimes for decades or more) before it occurred to anyone to attempt to reduce the one to the other. In such cases, reduction *primarily* involved an attempt to probe to a deeper level of physical ontology (which is precisely why *concept* replacement there worked so well). The reduced theory, however, may actually have been in little or no experimental or conceptual difficulty.

It should be said at this point that the direct *theory* replacements I am now considering are not "mere" or "total" replacements. In these cases, the reducing theory uses most or all of the same terms as does the reduced theory; also, the rearrangement of the terms is not totally unrecognizable. If they had been total replacements, one would not have been able to obtain in a reasonable way an approximation of one theory from the other. A *total* replacement, though perhaps possible "in principle," does not actually arise in the history of science. The kind of "reduction" envisioned by the analysis of Kemeny and Oppenheim—a reduction that requires no direct relation between the theories in question apart from their being related to the same data base (domain)—simply does not occur.

We are now in a position to understand why the notion of *approximation* plays an essential role in reductions of the direct theory replacement type. In these reductions, because B has run into trouble and is known to be *incorrect* in certain specific ways, A is *introduced* for the specific purpose of replacing B. But this means that A *must yield different results* from B for those areas in which B is incorrect. On the other hand, deductions from A must *agree* with those from B in those areas in which B is correct. This can be achieved (barring *total* replacement *à la* Kemeny and Oppenheim) only by A's involving in part at least (perhaps no more than) a *rearrangement* of the terms of B in its own laws (possibly with some new terms or significant new constants—such as Planck's constant h in quantum mechanics or the velocity of light c in relativistic mechanics) so that different deductive consequences may be drawn in some areas (sub-domains) and the *same* consequences in others. Only in this way can one generate, from theory A, an *approximation* of theory B for certain cases.

In the concept replacement reductions discussed earlier, on the other hand, reductions in which B is in no trouble at all, no occasion arises for the use of this notion of approximation.[11] In those cases, the intended theory replacements were *ontologically* motivated; the replacements (of *concepts*) there involved might have been viewed as identifications (from the "practical" point of view). Here, the replacements (of *theories*) are actually required, are required for reasons more mundane than those of ontology: the reduced theories are known to be *false* (although approximately correct under certain limiting conditions—therefore, quite useful from the *practical* point of view discussed in the last chapter).

Another characteristic of these direct theory replacement reductions, already mentioned but worth emphasizing, is that the replacing theory is essentially a theory at the same physical level—designed for the *same domain*—as the reduced or replaced theory; in the cases examined earlier, a deeper or more fundamental domain is invoked. Classical Newtonian mechanics is a theory which organizes our knowledge of motion and its causes without going to a deeper level. Special relativistic mechanics serves the *same function*. It is of course a different theory for the same domain—indeed, a radically different theory—but it does not replace the earlier theory by probing beneath their common domain; it simply organizes things differently within that domain. (Even a profound reorganization is a reorganization.)

Quantum theory is (in part) also a theory of motion and its causes replacing classical mechanics. It might be thought that quantum mechanics deals with a deeper level than that of classical mechanics. Such a view would, however, be superficial and misleading. It must be remembered that classical mechanics purports to apply to micro-phenomena. To be sure, classical mechanical results for atomic phenomena are quite wrong; but the theory does include atomic phenomena as part of its originally intended domain: nothing *in* the theory suggests that it might not apply at this level of matter.[12] It just happens to yield false predictions at the atomic level; in fact, it is just this *empirically* discovered fact which motivates the replacement of the theory.

It should also be noted that specific views as to the constitution of matter are not part of quantum mechanics any more than they are part of classical mechanics. "Matter theory" is quite distinct from each.[13] In principle, both classical and quantum mechanics could be applied to a Cartesian plenum as well as to the current view of the constitution of material objects. The same point regarding domains can be made about general relativity. It too reorganizes our knowledge of its domain without probing beneath it. The reorganization is deep and even revolutionary, but it is a reorganization nevertheless. Briefly, then, *none* of these *theory replacement reductions* are *microreductions*.

Let me sum up briefly, now, the characteristics of the kind of reduction under discussion. These are reductions in which:

(a) *B*, known to be false, is an *approximation* of *A*, in a clearly specifiable sense involving the taking of limits.

(b) *A* may be said to reduce (in Nickles' sense) to *B*.

(c) A *theory replacement* occurs via a *term rearrangement*.

(d) *A* and *B* apply to and explore the *same domain* (no microreduction).

Reductions of this type may be characterized as *domain preserving theory replacements*. By way of contrast, we may designate the kind of reduction discussed in earlier chapters as *domain eliminating concept replacements*. In reductions of that type:

(a) *B* is not in any asymptotic sense an approximation of *A* (although one may deduce approximations of the consequences of *B* from *A* for the observed phenomena common to *A* and *B*).

(b) It is not usually said that *A* reduces (in Nickles' sense) to *B*.

(c) The theory replacement is the result of an *explicit, item-by-item concept replacement*, which provides the focus of the reduction. *B*-items are replaced by *complexes* of *A*-items.

(d) The reducing theory *A* introduces a new domain—or explores an older, distinct domain—said to *underly* the domain of *B*, which it replaces. These are *microreductions*.[14]

Finally, let us review the relation between Nagel's notion of homogeneity and the distinctions I have already drawn. All concept replacement reductions are inhomogeneous. Conversely, all inhomogeneous reductions are concept replacement reductions. All theory replacement reductions are homogeneous; or rather, all *pure* theory replacement reductions are homogeneous, for there are reductions which share features of both concept replacement and theory replacement. (The general relativity/classical mechanics case is primarily a theory replacement reduction, but an element of concept replacement obtains there because of an element of inhomogeneity—the *force* concept in the reduced theory.) All homogeneous reductions are theory replacement reductions.

Reduction and Explanation Again

At the end of Chapter 4, I considered the relation between *concept replacement reduction* and explanation. Having now considered another kind of reduction, *direct theory replacement reduction*, it seems appropriate to consider the relation between explanation and this form of reduction as well.

Does special relativistic mechanics *explain* classical mechanics? In particular, is the basic law of motion of classical mechanics, $F = ma$ (or $F = d/dt(mv)$), explainable on the basis of Einstein's relativistic replacement of this law,

$$F = \frac{d}{dt} \left(\frac{mv}{\sqrt{1 - v^2/c^2}} \right) \quad ?$$

The answer, in my view, is No. There are several reasons for this, one implicit in the question itself. The relativistic force law is intended to *replace* its *defective* classical counterpart, not to explain it.[15]

We have already seen, in discussing direct theory replacement reductions, that '$F = ma$' is known to be *false*. If this law (or this law-like statement) is read as making the claim that forces cause accelerations proportional to them (with *mass* as the constant of proportionality), then it is simply false; it does not actually describe or refer to any feature of the world to be explained or accounted for. One would of course expect a putative explanation of $F = ma$ to answer the question "Why do forces produce accelerations proportional to them?" But the relativistic force law does not answer this question, for it implies instead that the statement assumed in the question is false. It is rather like asking "Why is $8+7=14$?"

Aside from the falsity of '$F = ma$', an answer to the question "Why do forces produce accelerations proportional to them?" would normally be made in terms of a deeper theory about the nature of force, mass, and acceleration. But that is definitely *not* what special relativistic mechanics provides. (*General* Relativity Theory provides that, because, as already pointed out, General Relativity has certain features of concept replacement reduction. Force *is* construed, in the *general* theory, as replaceable by a function of space-time curvature.) Special relativistic mechanics simply provides *another competing* force law, states a *different* relation between force and acceleration than the one assumed in the original question. It gives us what was earlier called a *rearrangement* of the terms of Newton's basic law of motion, not a deeper statement accounting for that law. (See pp. 123-125 of my book on relativity theory.)

Although it is false, '$F = ma$' is *approximately* true for velocities that are small relative to the velocity of light; the truth which it approximates *is* explained by Einstein's relativistic force law. In other words, *the relativistic force law explains the phenomena which the classical law almost explained*, and relativistic mechanics explains those phenomena which classical mechanics almost explained. But this does not mean that the former law explains the latter "law" (which is false), or that the former theory explains the latter theory.

According to the deductive nomological analysis of the concept of explanation, $F = ma$ *is* explained by its relativistic counterpart, because the former (or an approximation thereof) is *deducible* from the latter (when $v/c \to 0$). What this "approximative deduction" achieves, however, is not an *explanation of* $F = ma$, but rather part of the *explanation of why* $F = ma$ *was believed to explain correctly* the phenomena it was for such a long time thought to explain (that is, because we had been dealing with situations in which $v/c \approx 0$ and failed to notice the discrepancies).

In general, if theory B has been reduced to theory A in the sense of *direct theory replacement reduction*, then:

(1) A does not explain B (because B is known to be false).

(2) A explains the phenomena B purported to explain.

(3) A helps to account for (or explain) the fact that B was *believed* to explain the phenomena of the domain common to A and B.

(4) Not only does A *not* explain B, it *could not* explain B because A goes no deeper than B does.

Notes

1. Carl Hempel, "Reduction: Ontological and Linguistic Facets," *op. cit.*, p. 193.

2. See the reference to their paper in note 9 of Chapter 1. The point in question can be found on p. 313 of Brody's collection (*op. cit.*).

3. The details may be found in most books on the theory of relativity. In my own book (*op. cit.*), see p. 121 and p. 124. Incidentally, some delicacy is needed in order to state precisely what this notion of an *approximation* is; for almost anything "approximates" anything else if the concept is defined loosely enough.

4. Thomas Nickles, "Two Concepts of Intertheoretic Reduction," *The Journal of Philosophy*, Vol. LXX, No. 7, April 12, 1973. When I first read Nickles' paper I checked the appropriate sections of my own book on relativity theory and, to my surprise, found that I had indeed used the expression 'reduce to' in just the way he describes (on pp. 121 and 124). Nickles is to be congratulated for having identified the sig-

nificance of this standard usage, previously ignored or simply missed by philosophers writing about reduction.

5. Michael Audi, *The Interpretation of Quantum Mechanics*, Chicago, 1973, p. 16. Some care is required here regarding what has just been accomplished. For if the "laws of classical mechanics" are meant to include all known specific force laws, one might reasonably *deny* that the reduction in question has taken place. The law of gravitation (to mention one example only) is not deducible from the laws of quantum mechanics, as these are ordinarily understood. One must distinguish, in both classical and quantum theory, between what may, on the one hand, be called "the general laws of motion" and, on the other, further statements involving a specific matter theory—statements regarding the existence of objects of a certain nature plus specific force laws stating the manner of their interaction. (See also Chapter 4.) The claim that classical mechanics is reducible to quantum mechanics is meant to refer usually only to the treatment of the "general laws of motion."

6. Mario Bunge, *Foundations of Physics*, New York, 1967, p. 240.

7. For the distinction between special relativistic *mechanics* and the special *principle* of relativity (and certain related concepts), see pp. 129-131 of my book (*op. cit.*).

8. One could go on in this manner to discuss various types of reduction from the point of view of the general logic of relations. I shall not do so here, however, for although it might be of interest, it adds very little to an understanding of the concepts of reduction in physical science.

9. See Chapter 11 of my book on relativity. For a statement of the position on radical meaning change which I am criticizing, see Thomas Kuhn, *The Structure of Scientific Revolutions*, Chicago, 1962, Chapter 9.

10. See Audi's remarks on the concept of a particle, *op. cit.*, Chapter 2. He argues that the particle concept of classical mechanics (not just the word 'particle') is preserved in quantum mechanics.

11. I leave aside the problem of mathematical complexity in which approximations are introduced to simplify the required mathematical computations. Such approximations occur across the board in physical science—even in mundane applications within a theory, where reduction is not even at issue. In fact, they occur much more often in concept replacement reductions than they do in the kind of reduction now under consideration! (This is closely related to the fact that, usually, concept replacement reductions do not meet the deducibility condition of the standard analysis. This was shown in the previous two chapters.)

12. Similarly, velocities close to the velocity of light (and larger than the velocity of light!) are included in the intended domain of classical mechanics. There is within classical mechanics no way to draw any line between large and small velocities or between large and small bodies.

13. I have already remarked on this very important point in note 5 of this chapter and in Chapter 4. See, also, my book on relativity theory, p. 27; and G. Feinberg, "On What There May Be in the World," in S. Morgenbesser, P. Suppes, and M. White, *Philosophy, Science, and Method*, New York, 1969, pp. 152-164. Feinberg writes (p. 154): "The possibility of other objects than particles is contained in relativistic quantum theory because neither relativity nor quantum mechanics [nor classical mechanics] prescribe what objects do exist. . . . these theories do place restrictions on the properties of objects, and . . . tell about how to describe the behavior of objects that are known to exist, but neither requires that any of the possible objects that they describe are indeed to be found."

14. My distinction between these two types of reduction resembles Wimsatt's distinction between "interlevel" and "intralevel" reductions. See p. 219 of his paper, cited in note 11 of Chapter 4.

15. For similar remarks, see Lawrence Sklar, "Types of Intertheoretic Reduction," *British Journal for the Philosophy of Science*, Vol. 18, No. 2, August, 1967, pp. 109-124.

Chapter 6

Entity Reduction

The concept of reduction has been applied thus far to a number of distinct items. I have referred to the reduction of *theories*, *laws*, *concepts*, and *entities*. Our central topic has been the reduction of one *theory* to another. What I have called the standard analysis elucidates this notion primarily on the basis of the deduction of *laws*; whereas, for at least one important class of reductions, I have used instead the notion of the replacement of *concepts* and of the *entities* to which they refer. (In the previous chapter, of course, I analyzed another class of reductions in terms of the direct replacement of *theories*.) Our earlier shift, from the analysis of laws to the analysis of concepts and entities (which some may consider to be of minor significance—if not a misguided maneuver altogether) provides an understanding of other features of the scientific enterprise, of some interest and importance. These final chapters will explore a number of these. Here, we shall consider situations in which *entity reduction* rather than theory reduction serves as the primary analytic notion.

It may be plausibly claimed that, in what I earlier called *concept replacement reduction*—microreduction—*entity* rather than *theory* reduction is the fundamental notion. All earlier talk about laws, concepts, and theories would simply become then the conceptual or linguistic garb of basic claims about physical entities. I believe this view is fundamentally correct, although this is not the claim I have in mind here. My present claim is a somewhat different one (which, I believe, supports the claim just noted above). I wish, here, to consider cases in which there exists an accepted theory *B* having a well-articulated set of laws that describe the behavior of the entities dealt with in its domain; in which a reduction to a *deeper domain* obtains, regarding which there exists *no generally accepted reducing theory A*, no theory approaching the systematicity and degree of acceptance of theory *B*. That is, I wish to consider cases of reduction *without a reducing theory*.

Let me offer a specimen case. Physicists now believe that atomic nuclei are composed of so-called elementary particles, such as (but not exclusively) protons and neutrons. In this sense it is true and not misleading to assert that: *Atomic nuclei are reducible to elementary particles*. The manner in which atomic nuclei interact with their orbital electrons and the manner in which orbital electrons interact with each other and with the orbital electrons of other atoms is considered to be well-understood. These features are in fact accounted for by our theory *B*—those aspects of elementary quantum mechanics known in its major application as "quantum chemistry." But there is no fully articulated and generally accepted theory of elementary particles at present. There are various laws (or generalizations) in search of a unifying theory, various partial theories which seem to work for various subdomains, various models that are partially successful; but there is no single, fully accepted, overarching theory unifying all of the phenomena of this domain.

I submit that the standard analysis of reduction provides no way of characterizing this type of case even though, pre-analytically, something appears to be going on here that may fairly be characterized as *reduction*. Failure, here, is due to the fact that the standard analysis requires the existence of *two theories*, which must be analyzable as two *deductively unified sets of laws* regarding specific sets of entities, before one can even begin to talk of reducing one to the other. For example, in his paper "Reduction: Ontological and Linguistic Facets,"[1] Hempel speaks of ". . . the relations that obtain between the terms and laws of two theories when one of them is reduced to the other." Of course, defenders of the standard analysis might simply point out that they are interested only in the reduction of one *theory* to another *theory*. But such a response does not alter or affect the fact that there appear to be cases of *reduction* with which the standard analysis cannot begin to deal.

I believe that my own view of concept replacement analysis fares better in such cases. This is partly because I have not required anything approaching deductive unity for the deeper theory—a step toward not having adopted a particular deeper theory.[2]   But it is only one step; and so, concept replacement analysis would appear similarly to be not quite suited to this type of case. For, *if* reduction is always *of a theory* and *to a theory*, if reduction is analyzed in terms of concept replacement (so that the laws of theory *B* transform into the laws of theory *A*), then our analysis is in trouble simply because there *is* no "theory *A*" to state the laws of. We should, however, be in a somewhat better position if we spoke directly of the *laws* (and entities) of the *deeper domain*.   After all, we do have some laws regarding the behavior of elementary particles, though they are not embedded in any overarching theory.   In this sense, concept replacement analysis, focusing as it does on deeper level *laws*, goes some way toward capturing cases of the sort in question.

The situation is even better when one speaks of *entities* directly. For it is possible to speak of a *set of entities* for which we have *no accepted theory*. This arises in a number of ways. Here are three kinds of pertinent situations.

(1) There may be several distinct *successive* theories about the nature of a kind of entity or domain: the series of such theories may be of such a sort that we have good reason to believe that the currently favored theory is not the final one—or anywhere near the final version. (We have learned much, and we know that there is still more specifically to learn.)

(2)   At a *given time*, we may have several *competing* theories of the entity or domain in question, each of which is held by its supporters to be "the" correct theory.

(3) At a *given time*, we may have several *non*-competing theories each of which, though *admittedly* not final, is able to handle its own *sub*-domain or certain restricted aspects of the entity in question—which other such theories cannot handle; thus we may have several partial theories or "models."

Sometimes, situations arise in which a combination of the above may characterize the actual state of research.

"Theoretical concepts" are *not* as totally theory dependent as certain recent philosophies would have us believe.[3] They are certainly not tied to any one specific theory which happens to employ them. For example, there have been a number of related *theories of the electron*, from Lorentz's through more recent quantum mechanical conceptions.

These differ non-trivially from one another, but they all purport to describe the same *entity*-type—they are all about *electrons*. "That about which" all of these theories have something to say may be characterized neutrally with respect to the conceptual apparatus or beliefs *specific to* any one of these particular theories. For instance, electrons may be identified as the elementary long-lived negatively charged constituents of ordinary matter, which exist as part of each atom of ordinary matter. These are the entities that each succeeding theory attempts to characterize, in developing detail, with greater accuracy, and through relations with other elementary constituents of matter. Here, we speak directly of an *entity* for which we have several *distinct theories*, in the manner of (1) above.

It may be argued that the description I have given is itself a "theory"—one so carefully stated as to straddle the fence with regard to the specific claims of more specific theories, but a theory nonetheless. I could grant this (although, I believe, I need not—the description given of electrons is patently *not* a "theory") without yielding my principal point. For I wish to stress that one may describe, clearly describe, what it is any specific theory of elementary particles is intended to be a theory *of*—and what any such theory ought in general to accomplish with respect to such entities—without having any specific such theory in hand. Here, I think we are speaking "directly of entities" rather than speaking of them only through the mediation of specific theories about them.

The failure to realize that this sort of situation arises (indeed, is hardly rare) is due in large measure, I believe, to the influence of certain radically contextualist theories of meaning—those of Braithwaite, Feyerabend, and Kuhn, for instance, referred to in note 3. I suggest further that a number of issues in the philosophy of science are likelier to be resolved if fewer attempts were made to tie them to a "theory of meaning." In my view, Feyerabend's earlier replacement analysis of reduction, for example, leans too heavily on a particular theory of meaning that is itself severely criticized. In consequence, his analysis of reduction suffers. My own replacement analysis does not similarly depend on a particular theory of meaning either for its motivation or its character.

In general, one need not subscribe to a comprehensive theory of meaning in order properly to understand the nature of scientific methodology. If a favorable analysis of an element of scientific methodology happened to accord with an already generally accepted theory of meaning, the congruence between the two might give that much more credence to the latter; but then, perhaps, the standing of the theory of meaning benefits from its being in accord with findings regarding the methodology of science!

A theory of meaning ought to do many things, but it ought at least to accord with scientific practice. Indeed (to borrow a well-turned phrase from Wimsatt, Chapter 4, p. 26), perhaps ". . . we must use our current and future intuitions and theories of [scientific methodology] to help develop our [theories of meaning]." My attitude does not preclude *changes* in scientific practice or its philosophic analysis, in the light of arguments based on a reasonably well-grounded and generally accepted theory of meaning—based even on well-grounded *fragments* of a theory where no general theory is forthcoming. But it is not incumbent on anyone attempting to analyze scientific methodology to *provide* initially and to defend a theory of meaning said somehow to "ground" methodology. Here, the issues may be fruitfully separated.

To return to the point of our discussion: if it makes sense to speak of *entities* without having accepted any specific theory about their detailed nature and behavior, then we may affirm that *atomic nuclei are reducible to elementary particles*, even though we lack a specific theory of the latter in terms of which to state some detailed *theory* reduction. (The appropriate reduction, here, would be of the concept replacement variety. If "general" concepts such as the fence-straddling characterization of the electron were allowed, then concept replacement reduction would obtain here too—but it would not be one in which specific laws transform into other specific laws.)

To put the matter another way, we know that *entities* of one type are reducible to those of another, even though we do not know how the correct *theory* reduction goes in detail. Notice: it is not the case that we merely *believe* that atomic nuclei will *eventually* be reduced to elementary particles: we *now know* that atomic nuclei are composed of elementary particles; we do not know just exactly how they are. The "exactly how" concerns the *theory* reduction (via a detailed concept replacement) that we do not have as yet.

So, in advance of having a specific, articulated, accepted theory of elementary particles, physicists now can quite correctly claim that atomic nuclei are composed of elementary particles (in some manner or other), that atomic nuclei are at bottom nothing but certain interacting combinations of these particles (as yet only partially understood), and that they are in this sense reducible to them.

An earlier finding supports this point. Since *concept replacement theory reductions* do not require the deduction of laws within the deeper theory, we can speak meaningfully of reduction (of the reduction of *entities*) even in the absence of an accepted deeper theory. Moreover, it is the notion of *replacement* itself, already developed for two kinds of *theory* reduction, which provides the basis for the idea of *entity* reduction here defended. For the concept of entity reduction can be elucidated further by the concept of replacement. To accept the claim that atomic nuclei are *reducible* to elementary particles is to say that one's physical *ontology* need no longer include atomic nuclei as fundamental constituents. They may *now be replaced*—from the theoretical point of view—for general ontological purposes, by interacting combinations of elementary particles. Yet the exact nature of these combinations is a function of some future *theory reduction* still to be determined by physicists.

The situation is similar to that discussed in Chapter 3, where it was argued that law deduction is not a necessary condition of reducibility of theories. There, however, we acknowledged an accepted reducing theory *A*. Relevant cases of that sort (for example, the thermodynamics/classical mechanics case (see Chapter 3) and the case under consideration here form the opposite ends of a spectrum of cases; they are not really distinct species of cases. The vision of one trouble-free, overarching, generally accepted deeper theory of a domain, from which all the laws of any reduced theory can be deduced with complete rigor, is the vision of an ideal of the physical sciences that, clearly, is almost never attained. The thermodynamics/classical mechanics case approaches the ideal more completely than the atomic physics case explored in this chapter, but they remain, nevertheless, elements along the same spectrum. Just as the lack of ideality in the former cases does not prevent us from claiming, correctly, that gases are composed of molecules, so too, here, we may claim equally correctly that atomic nuclei are composed of elementary particles and thus are reducible to them. It is even possible that cases of entity reduction arise in which no single accepted *reduced* theory obtains. For example, if the quark hypothesis became well-confirmed and generally accepted, we could say that such "elementary particles" as protons and neutrons have been reduced to quarks even though we have no fully articulated theory of the former. Two considerations, then, have facilitated our argument: the shift from laws to concepts in fixing the focus of certain cases of reduction; and the realization that law deduction is not a necessary condition for reduction. Further advantages will be explored in the next chapter.

Notes

1. *Op. cit.*, pp. 179-180.
2. I have a further comment to make about this at the end of the chapter.
3. Perhaps the classic contemporary statement of the view that theoretical concepts are totally theory dependent may be found in R. B. Braithwaite, *Scientific Explanation*, Cambridge, 1953. It has subsequently appeared in a variety of forms and in the writings of philosophers of different persuasions (e.g., T. S. Kuhn, *op. cit.*). One of its most influential recent proponents is Paul Feyerabend. At least, he held the view when he wrote "Explanation, Reduction, and Empiricism," in H. Feigl and G. Maxwell (eds.), *Minnesota Studies in the Philosophy of Science*, Vol. 3, Minneapolis, 1962. N. R. Hanson also held this position in his book, *Patterns of Discovery*, Cambridge, 1958. I have criticized certain aspects of the general thesis in "Models and Theories," *British Journal for the Philosophy of Science*, August 1965 (reprinted in Brody's collection, *op. cit.*), and in Chapter 11 of my book on relativity theory (*op. cit.*). An incisive criticism of aspects of the position may be found in Peter Achinstein's "On the Meaning of Scientific Terms," *The Journal of Philosophy*, Vol. 61, September 1964, pp. 497-509. Indeed, Achinstein's paper ought to make one suspicious of any radical views regarding the nature of science based on general theories of meaning. In my view, the lesson of his paper is this: that sensitivity to the complexities of actual scientific usage will tell one much more than a general theory of meaning imported from another domain.

Chapter 7

Branch Reduction

In these last two chapters, I shall try, with the aid of my concept replacement analysis of one type of theory reduction, to clarify certain further aspects of the reductive enterprise in science. My purpose is two-fold. First, I wish to discuss an aspect of the reductive enterprise which has received less attention than it deserves. Second, I intend thereby to illustrate a further application of the central idea of concept replacement.

It is generally recognized, today, that *chemistry is reducible to physics*. The reduction differs, however, from all other examples so far considered, in that 'physics' and 'chemistry' do not refer to specific theories (or laws, or entities) but to scientific disciplines to entire *branches of science*. I have talked so far of *theory* reduction, *concept* reduction, *law* reduction, and *entity* reduction; now we must consider the possibility of reducing one *branch of science* to another. I shall examine in some detail the branch reduction of chemistry to physics. Much of what may be said here can be generalized to other cases (for example, to the case of biology/physico-chemistry); but each such case has its own distinct and important characteristics.

One might favor the following strategy for analyzing the concept of branch reduction: break up each "branch of science" into its constituent theories; then show how each *theory* of the reduced branch (possibly all of the theories) is reducible to theories of the reducing branch (possibly some single theory). This is essentially the course followed by Kemeny and Oppenheim, in the paper of theirs cited a

93

number of times. It seems also to be the course followed implicitly by Hempel, in the paper of his mentioned earlier on; for Hempel considers the possible reduction of biology to physico-chemistry, viewed as a case of *theory* reduction, and he applies to it his analysis of inter-*theoretic* reduction.

In the case under consideration, however—if not in all cases of branch reduction—that procedure is unnatural and misleading. Indeed, it yields incorrect results; for, as a branch of science, chemistry is not characterizable simply as a conjunction (or set) of specific theories. Of course, specific theories are part of what is involved in the science, but they are only part. Moreover, there is no neat, brief, general, or informative way of characterizing "The Science of Chemistry" so as to include all and only those problems, theories, concepts, procedures, domains, and equipment which are dealt with, accepted, followed, examined, or used in university "chemistry departments," at research centers, or at manufacturing companies. The agglomeration of those items presently included under the heading 'chemistry' has been shaped by a large number of factors, both historical and contemporary. They include for instance such diverse items as metaphysical ideas about the kinds of thing populating the universe, personal interests and professional inclinations of individual "chemists" (or even of physicists or biologists or engineers), the specific kinds of instrumentation available at given times, alternative theories (certainly theories), models, laws, generalizations, presuppositions (hunches?) accumulating around particular questions, problems, or phenomena believed for one contingent reason or another to be especially significant or interesting; they include such items, I submit, *plus* what might be called a form of "institutional inertia." Professionals—chemists, physicists, engineers, and the rest—deal with the problems they do and in the way in which they do, because they have learned that they are *supposed to* do so, learned it in school or by observing their colleagues.

If this account is correct, or even roughly correct, then the reduction of chemistry to physics may be more appropriately analyzed as the reduction (in some relevant sense—or perhaps as involving some other relation or relations) of each of the diverse kinds of items mentioned to comparable items in the science of physics. To show in detail how to do this would be an enormous undertaking. Indeed, *if* each and every item in the list offered above were considered equally significant, then the reduction of chemistry to physics has *not* yet taken place! Chemistry departments still exist, after all, and rightly so. Even if it is the case, as I should argue, that in matters of *fundamental physical ontology* (that from the theoretical point of view discussed earlier) chemistry has indeed been fully reduced to current microphysical theory, it remains true that the characteristic concepts, laws, instruments, laboratory procedures, etc. of chemistry are at least convenient and perhaps indispensable for many or most of those concerns that exhibit what I termed the "practical attitude." (Even theories known to be false, after all, may be useful for many purposes, for example, classical Newtonian mechanics.)

Let us therefore, in pursuing our question, put aside as far as possible those practical items that seem to be of less importance in characterizing a scientific discipline (for instance, instrumentation, laboratory procedures, etc.). Let us focus rather on the characteristic *subject matter* or *domain* of these branches of science. Any attempt to characterize chemistry and physics in this way will still be problematic, for it will involve a large degree of *overlap*. But one may characterize their distinction roughly by attending to the "center of gravity" of their respective domains. I shall press the point—but only on the understanding that what follows is doubly idealized: for I am waiving all considerations but those of subject matter, and the distinction I intend to draw in terms of subject matter is not as sharp as I shall make it seem. (I shall have more to say, later and in the next chapter, about the matter of overlap—which I think has great significance.)

Roughly speaking, then, relative to physics viewed as a distinct science, and relative to their recent history, chemistry has traditionally dealt with grosser features of matter. (Even this needs qualification, for clearly astronomy is not part of chemistry but of physics. Or, ought astronomy to be considered a branch of science distinct from both? By 'physics' I have in mind certain aspects of what I referred to as "matter theory" in earlier chapters.) Traditionally, chemistry has been concerned with: (a) the properties and interactions of various gross *kinds of substances* and the network (or heap) of laws and generalizations discovered at this level (*descriptive chemistry*); and (b) attempts to explain the foregoing on the basis of theories about the constitution of, and about the behavior of the constituents of, those substances (*theoretical chemistry*). These two aspects of chemistry may be recognized in the chapter headings of the usual chemistry textbooks. The former is illustrated by such titles as "The Halogens," "Phosphorus, Arsenic, Antimony, and Bismuth," and "Water"; the latter appears in such chapter headings as "Covalence and Electronic Structure," "Oxidation-Reduction Reactions," and "Complex Ions."[1] It is theoretical chemistry, the second aspect of the traditional science of chemistry, that involves a large overlap with (aspects of) physics. To put the point in a word, both chemistry and physics deal with the structure of matter at the level of atoms and molecules.

Even within the area of overlap, however, there remains a significant *difference* between theoretical chemistry and what might be called *atomic and molecular physics*. It is this. The *interest* and *intent* of the theoretical *chemist* lies in determining the molecular (or atomic) arrangement of the sorts of substances that he deals with in (or at the level of) descriptive chemistry; with that information, he may explain (or predict) certain favored features of the behavior of those substances, from *given* pertinent laws. On the other hand, the atomic or molecular *physicist* is primarily interested in *accounting for* the arrangement and behavior of atoms and molecules; he does so on the basis of current physical theories regarding the structure of the atom and the laws governing the behavior of its constituents—that is, on the basis of quantum mechanics.

We may then say that the concepts and laws involved in *microphysics*—an atomic theory (a "matter theory") plus quantum mechanics[2]—are used to *explain* what is the *starting point* for the theoretical chemist. The area of overlap includes both the *basis for explanations* for the chemist and the *problem area* for the physicist. For example, the chemist will *use* the concept of *covalence* as a "primitive" to help account for the bonding characteristics of certain kinds of atoms and for certain features of gross kinds of substances and their interaction, whereas the physicist will account for the occurrence of covalent bonds on the basis of more fundamental quantum mechanical concepts. (In particular, the *chemist* will say, e.g., that molten sulfur is an insulator because the S_8 molecule is formed from covalent bonds, which are very strong and so do not form the ions needed for conduction. The *physicist* will be interested in pointing out that covalent bonds occur in general because of certain features of the quantum mechanical wave function Ψ for such systems.)

This characterization of the difference between chemistry and physics (with respect to subject matter) is neither new nor controversial. The distinction is already recognized by scientists working in these areas. ·Let me, however, offer one long but typical quotation to support this claim. In Wolfgang Finkelnburg's book *The Structure of Matter* (New York, 1964), in a section entitled "Molecular Physics and its Relation to Chemistry" (pp. 343-344), we find:

Molecular physics is the science of the structure and properties of molecules so far as they are determined by physical methods. In this sense it is a logical continuation and extention of atomic physics proper. From what has been said it is obvious that molecular physics is closely related to chemistry. Chemistry attempts to determine the composition of a compound and the formula of its structure (e.g., of CH_3Cl). From such chemical formulae, the chemist derives conclusions as to the behavior of molecules, i.e., the possibility of its reaction with other molecules or atoms. The correctness of the chemical formulae is checked by chemical analysis [involving laboratory procedures mentioned earlier] and finally methods are developed for synthesizing the molecules from their constituent elements. Chemical methods are also used for determining molecular properties and certain characteristic quantities, among them the heat of formation of the molecule, e.g., the energy freed in forming HCl from 1/2 mole H_2 and 1/2 mole Cl_2 [derivation of laws of descriptive chemistry]. Chemistry is not able however . . . to *explain* the chemical valency of the atoms forming molecules, nor the molecular structures found, nor finally the different stabilities of different bonds in molecules.

It is therefore the most important task of molecular physics to determine how the atoms are bound in a particular molecule. The young student of chemistry often asks himself the question why there is a stable molecule NH_3 but not NH, a stable CO and CO_2 but no CO_4. It is an achievement of atomic physics with its quantum mechanical atomic theory that we now have a basis for answering this fundamental question by a theory of the chemical bond. It is, therefore, not an overstatement when we say that only atomic and molecular physics has furnished a solid theoretical foundation for its older sister science, chemistry. . . . The chemical behavior of a molecule should follow from the correct physical picture of the molecule, and it is used to verify the correctness of this picture.

It must be emphasized that there is a great overlap here—an overlap in part due to the overwhelming success of the reduction. For example, physicists, with the aid of computers, can apply quantum mechanical concepts and laws to deduce not only the orbital electronic arrange- ment of single atoms and of simple molecules (the limits of what could be done before the advent of computers), but they can also encroach upon the traditional territory of theoretical and descriptive chemists by deriving statements describing the behavior of more complex molecules, reaction rates, etc. (Or should we say that chemistry is encroaching upon physics if it is a "chemist" by training and departmental affilia- tion who uses and refines quantum mechanical techniques to account for some of these "chemical phenomena"? Is Linus Pauling a chemist or a physicist?) The overlap involved distinguishes this case of reduc- tion from all others so far considered. To quote from Finkelnburg again (p. 344):

> The results of three decades of research in molecular physics are so striking that modern inorganic and physical chemistry to an in- creasing extent not only has taken over many individual methods of molecular physics, but is actually doing molecular-physics research just as it was done before by physicists. There is scarcely a border line any more between physics and chemistry. . . .

The significance of the overlap will be considered further in Chapter 8.

Now, how is this example of a *branch reduction* to be characterized? How can we apply the analytical machinery developed in earlier chap- ters? I believe the following somewhat complex description to be necessary.

First, chemical *concepts* (valence, bonding, catalysis, . . .) may be *replaced* one-by-one with complex constructions of micro-physical concepts (concepts from Finkelnburg's "atomic physics with its quan- tum mechanical atomic theory") in such a way that chemical laws may be transformed into statements of quantum mechanics plus atomic theory (micro-physics) which can be shown to be true—either by deduction or in some other way.[3] So far, then, I propose that *concept replacement* analysis, developed for domain eliminating *theory* reductions (microreductions), applies as well to this case of *branch* reduction. The relevant concept replacements have taken place gradually over a period of time; the reduction has been an ongoing process. Hence, it makes sense to speak of the progress of the reduction on the basis of how many chemical concepts have been gradually replaced, how much evidence has accumulated for the correctness of each proffered replacement function, etc. (See the last quotation from Finkelnburg.) In Chapter 5, on the other hand, we saw that such considerations (involving *gradual progress* in a reductive venture) make no sense at all with regard to domain preserving direct theory replacement reductions—the case of relativistic mechanics/classical mechanics, for instance. This kinship between concept replacement *theory* reductions and the reduction of one *branch* of science to another should come really as no surprise. The former, of course, involves the explicit introduction (or use) of a *different domain* to account for the phenomena of the reduced theory; similarly, different branches of science are normally thought to involve different domains. More will be said about this connection, below, and in the next chapter.

Second, we now know that the *entities* dealt with by descriptive chemistry (ordinary "sorts of substances") are *composed of* entities dealt with by micro-physics—atomic theory plus quantum mechanics. This was already known before quantum mechanics reached its present status as the accepted theory of micro-physical phenomena. That is, an appropriate *entity reduction* (discussed in Chapter 6) was acknowledged to obtain even before the detailed development of a fully acceptable reducing *theory*. So certain features of what I have called *entity reduction* are also to be found in our case of branch reduction. Indeed, even today, very few of the more highly complex laws of descriptive chemistry can be *deduced* in detail from *currently* accepted microphysical theories. It is generally thought that the shortcoming is due primarily to the formal mathematical complexity of applying quantum mechanics to the gross phenomena of descriptive chemistry. No one, however, worries that chemical substances are *not* actually composed of physical atoms. Here, one may recall the notorious complexities of organic chemistry. Before modern computers could relieve the sheer drudgery of the branch reduction we are considering, the following remark, written in 1944, in a book titled *Quantum Chemistry* (*sic*), was entirely commonplace:

> In so far as quantum mechanics is correct, chemical questions are questions in applied mathematics. ... No chemist ... can afford to be uninformed of a theory [quantum mechanics] which systematizes all of chemistry even though mathematical complexity often puts exact numerical results beyond his immediate reach.[4]

The reduction of chemistry to physics is thus typically admitted in the face of non-deducibility, which conforms with my general claim (Chapter 3) that law deducibility (condition (i) of the standard analysis) is not a necessary condition for reduction. Three decades of improved computer technology have provided the chemist more "exact numerical results"; but the detailed deducibility from fundamental micro-physical laws of the entire range of phenomena of descriptive chemistry remains an as yet unattained ideal.

Third, the *ontological implications* of concept replacement analysis (explored in Chapter 4) are also pertinent in our chemistry/physics case (and in other cases of branch reduction). The continued success of an ongoing branch reduction, evidenced by more numerous concept replacements, signifies the step-by-step replacement of one *ontology* by another. We now know that the entities dealt with by descriptive chemistry are composed of entities dealt with by micro-physics (atomic theory plus quantum mechanics). But this means that, for general ontological purposes (in the sense described in Chapter 4), we need not, in describing what is taking place in any pre-analytically identified *chemical domain*, supplement the conceptual apparatus of microphysics with any specifically chemical concepts or laws. Of course, for practical purposes, as already noted, the conceptual apparatus of chemistry (theoretical and descriptive) may be quite convenient and even indispensable. (Non-deducibility alone would insure this.) The chemist who admits that quantum mechanics "systematizes all of chemistry" need not, for that reason, vote for the abolition of his university's chemistry department. Indeed, the point raised by Rosen and Pattee (Chapter 4) regarding the relation between thermodynamics and classical mechanics may be made with equal force for the chemistry/physics case. Perhaps we must (paraphrasing Rosen and Pattee) already have the *chemical* laws before it dawns on us that we should attempt to deduce those laws from basic micro*physical* laws. So seen, their point argues a special application of a more general practice (see also, Wimsatt's related point, quoted in Chapter 4). But the most that this would show would be the *importance* of the practical attitude— which does not affect our present issue.

These three considerations, then, bear on the reduction of one branch of science to another (or rather, bear on those aspects of the reduction concerned directly with the *subject matter* or *domains* involved). I believe that concept replacement analysis applies to other instances of branch reduction, but I hesitate to claim that all branch reductions have all and only those features mentioned in the chemistry/ physics case. As remarked earlier, each case of branch reduction is bound to have its own distinguishing features; these *may* prove diverse enough to disallow any quick generalization from the chemistry/physics case alone. But there is an advantage in attempting to analyze other instances of branch reduction (successful ones, unsuccessful ones, those not *yet* successful or subject to controversy), by considering the extent to which the analysis provided is paralleled in these other cases. I suggest for instance that the issue of whether (how, or to what extent) *biology* may be reduced to *physics and chemistry* is best pursued by way of the following questions:

To what extent can specifically biological *concepts* be replaced by constructions out of chemical and physical *concepts*? Can biological laws be transformed into physical and chemical laws? Is biology "simply" applied physics and chemistry?

To what extent can we now claim that biological *entities* are wholly composed of physical and chemical *entities*, lacking as we do detailed law deductions?

Do we need specifically biological concepts or are physical and chemical concepts sufficient for formulating what there is at bottom in the physical world?

To what extent is there a significant overlap between the domains of biology and physics-chemistry?

Answers to these questions would considerably clarify the reducibility of biology to chemistry and physics.[5] One particular biological concept, however, is thought to challenge the venture. The concept of *function* (or more loosely, of *purpose*) is said not to be reducible to physical and chemical concepts—to distinguish, for that reason, biology from the physics/chemistry case. Unlike other "biological concepts," *function* does not refer to an element or a group of elements within the *domain* of biology, but only to a way of viewing selected relationships among those elements or the elements of any other domain. The question whether "functional explanations" in biology can be "reduced to" explanations not involving functional notions is a critical part of the question whether biology is reducible to chemistry and physics. I shall not concern myself here with the issue of *functional explanations*. In my view, their "reducibility" to strictly causal explanations is still an open question.[6] I wish to point out only that if the concept of function in biology were as distinctive as it is claimed to be, it would serve *methodologically* to sort that branch of science from physics and chemistry and to qualify decisively the reducibility of the first to the second.

This clarifies, incidentally, what I meant by saying that branch reductions may differ from one another in certain important ways; for the physics/chemistry case does not involve any "methodological concepts" (like that of *function*) that are thought to be essential in the one branch and not in the other.

One final observation regarding the physics/chemistry case is in order. In reducing chemistry to physics, a *variety* of sub-domains, laws, models, theories, . . . (of chemistry) are reduced to *one* physical theory (what I've called *microphysics*: quantum mechanics plus atomic theory). Here, the reduced branch (of our branch reduction) involves no known, clearly formulable unifying theory though the reducing branch does. Hence, the reduction brings about a type of *unification* not previously provided. To put the point another way: insofar as anything may be called "*the* theory of chemistry," it is a *physical* theory—quantum mechanics (plus atomic theory). The various sub-domains and laws of chemistry achieve their unity only at the deeper level of micro*physics*. My conjecture is true as well of other cases of branch reduction. For example, we may anticipate that various biological sub-domains, concepts, and laws will be theoretically unified only at the deeper level of physical and chemical concepts and laws. Perhaps the relation between psychology and biology (including neurophysiology) will be found to be similar in this respect. But each such case of branch reduction would have independently to be examined in its own turn. Our speculations must accommodate the contingencies of each science.

Notes

1. These are taken from a textbook which Linus Pauling wrote, called *General Chemistry*, San Francisco, 1947.

2. See Chapter 5. In the remainder of this chapter, I will use the term 'micro-physics' to refer to the conjunction of these two items.

3. See Chapter 3, where I discuss the fact that law deduction is not a necessary condition of reducibility.

4. Eyring, Walter, and Kimball, *Quantum Chemistry*, New York, 1944, p. iii. This quotation is from the opening two sentences of the preface.

5. In my opinion, on the basis of such answers, this reduction is proceeding quite well. I shall not defend the claim here, however.

6. It would be interesting to investigate the form of *reduction* involved in the claim that functional explanations are "reducible to" causal explanations. I suspect that it is concept replacement (the concept of *function*) that is methodologically critical here.

Chapter 8

The V_a/V_b Vocabulary Distinction

Only *one domain* is involved in the chemistry/physics example of branch reduction. Chemistry and physics exhibit no natural boundaries providing a sharp, non-arbitrarily defined division between two theoretically separate domains—"the Physical" and "the Chemical." The overlap stressed in the preceding chapter is not merely a fuzzy border area: the two "branches of science" shade into each other in a clearly specifiable manner. It would not be misleading for instance to say that chemistry is a branch of physics! This would be a natural way of reading the remark of Eyring, Walter, and Kimball (quoted in Chapter 7) to the effect that ". . . chemical questions are questions in applied mathematics." Problems in theoretical chemistry may be fairly viewed as problems in the *application* of quantum mechanical laws to the complex atomic and molecular systems that we know, at bottom, "chemical entities" to be.

The two domains of interest—the chemical and the physical—are distinguished best *historically*, largely by reference to features I introduced in the previous chapter, which I explicitly prescinded from the discussion: instrumentation technology, laboratory procedures, the interests and education of influential physicists and chemists and other scientists, and similar factors there mentioned. It is in large measure due to hindsight born of the success of this branch reduction that we know this relationship to hold, but it holds nonetheless.

But if we concede that it was an historical accident to think in terms of two *theoretically distinct* domains, we need no longer preserve the vestiges of the original "error"; we need no longer employ characterizations that suggest a deeper distinction and that may generate deeper philosophic problems as well. I have in mind particularly the philosophical puzzles' that arise from proceeding straightforwardly on the basis of positing two distinct *vocabularies*, the "physical" and the "chemical" ('V_p' and 'V_c', as they might be termed, following the conventions that have been used throughout this account).[1] Indeed, much of what has been written about reduction has been obviously handicapped by too great a stress on a *general V_a/V_b* distinction. Such a distinction for instance is regularly employed in the standard analysis, and I have myself been using it without question until now. Most current discussions of reduction *begin* by drawing the distinction. And that very way of proceeding automatically (and almost unnoticed) generates a serious conceptual problem for ensuing discussions: for how are the *distinct* vocabularies (and thus the distinct *concepts*) of the domains in question to be *related* to each other—where B is said to be reduced to A (where 'A' and 'B' refer to specific *theories* or entire *branches* of science)?

Here, I should like to review the proper role of a V_a/V_b vocabulary distinction.[2] I should like to consider what follows from the fact that the chemistry/physics case involves a spectrum of related phenomena rather than a sharp division of domains. I begin by introducing and defending what may at first appear a rather bizarre claim about the chemistry/physics example of branch reduction.

From the point of view of physics (i.e., *microphysical theory*[3] —a view regarding the atomic constitution of matter plus the quantum mechanical laws governing those constituents), would-be *chemical terms* (our V_b) may be rightly construed as *physical terms* used to refer to various items (things, properties, processes, etc.) not yet fully understood on the basis of microphysical theory, or, alternatively (when those items become sufficiently understood) as shorthand for the complex physical representation of them. This construal may appear quite *post hoc*, since it presumes the eventual, successful reduction of "chemical concepts" and theories to "physical concepts" and theories. Actually, it says that there are no essentially irreducible chemical terms or concepts, that such terms and concepts are only physical terms and concepts not (yet) replaceable by constructions out of more basic ones. There remains *no distinction of two vocabularies* said to be charac-

teristic of two branches of science having two *subject matters*; there
is only the less exciting (but more enlightening) distinction between
more complex *physical* entities, processes, etc., partially but not yet
fully accounted for on the basis of fundamental microphysical concepts
and laws, and those other physical entities, processes, etc. which *are*
fully accounted for on this basis.

The proposal's apparent strangeness dissolves when two things are
recalled. First, our way of construing the V_c/V_p "distinction" simply
follows from the kind of claim made by Eyring, Walter, and Kimball. It
is merely the result of taking seriously the idea that chemical problems
are problems in applied mathematics (the complex problems of applying basic quantum mechanical laws to those physically complex systems
called chemical systems). It conforms, also, with the recognition by
scientists of the great overlap between the two domains (see the quo-
tations from Finkelnburg, in Chapter 7).

Secondly (and more important), physical theory itself is not as fully
and successfully deductively unified as the writings of philosophers lead
one to believe. The point has already been considered in the context of
discussing the deducibility condition which the standard analysis insists
upon. (In Chapter 3, for instance, I showed how the requirement is too
strong, as it would eliminate accepted cases of reduction.) The crux is
this: physics *already contains* and deals with laws, concepts, processes,
etc. not yet fully understood on the basis of its own fundamental concepts and laws (or of any reasonable set of such "basics"). But such
laws, concepts, etc. are not, *for that reason*, construed to be part of
another branch of science. There is usually a spectrum of such cases to
be considered; the idea of being only "*partially* understood" is familiar
to workers in all areas of science. Full deducibility is but an ideal in
most areas of science.

Now, if the situation exists *within* what everyone recognizes to be
physics, then the suggestion is simply that chemistry be viewed as part
of physics *along the same spectrum* of cases. It is of course a part that
happens to have its own distinctive *history*, but it is not, for that reason,
any the less a part of physics. (In fact, with regard to our understanding
phenomena on the basis of physical foundations, the "chemical" part
of this spectrum clearly overlaps portions of the "physical" part.)

I trust that the suggested construal seems more natural now. Consider the "chemical term" 'acid'. I claim that, even before acids were

fully understood on the basis of microphysical theory, one could reasonably have viewed 'acid' as a *physical* term, that is, as designating a complex physical substance displaying a characteristic molar behavior the microstructure of which was not yet *fully* known or understood. The *concept replacement* was, therefore, already accomplished in outline; only the details remained to be worked out. (I.e., it was *known* that acids had a particular *physical* microstructure, that they were composed of atoms having certain sorts of arrangements; but the detailed behavior of those arrangements was not yet fully understood.) Once achieved, the term 'acid' could be read as shorthand for the complex physical expression (construction) designating the structure in question. Such terminological and conceptual shifts regularly occur in science. (See Chapter 4, for the ontological implications involved.) Thus, the V_a/V_b distinction may be viewed as part of a *understood/not-yet-fully-understood* "distinction" within the reducing branch—the "*A*-science." Since the *A*-science has its own terms referring to other not-yet-fully-understood items within its own domain, the difference between these and V_b referents is one of degree only. Indeed, *at best* one of degree only, for certain areas of physics are less well understood quantum mechanically than certain traditional areas of chemistry.

The overview I am advocating has a number of advantages. First, it accounts conveniently for the overlap of *A* and *B* with regard to branch reduction. Certain concepts are both "chemical primitives" and physical concepts in need of quantum mechanical analysis. If we look at the terminological situation thus, we can also clarify certain aspects of research in the area of overlap. Attempts to explain "chemical phenomena" on the basis of the conceptual apparatus of quantum mechanics may be viewed as a continuing process of making *concept replacement reductions* or, alternatively, as attempts to understand fully not-yet fully-understood *physical* entities. It should now be clear that these are just *two ways of viewing the same endeavor*. As I claimed earlier,[4] there is no significant theoretical difference between saying, "*B* reduces to *A*" and saying, "*B* is a branch of *A*."

Hempel has written that labels like 'physical', 'chemical', 'biological', etc., ". . . apply properly not to particular objects or events but to particular ways of characterizing them. The same remark refers also to *kinds* or *classes* of things or events."[5] Thus, a given substance may have both a *biological* and a *physico-chemical* characterization, depending on whether we use biological terms or physico-chemical terms to describe it. Hempel considers the substance *epinephrine* and speaks of its biological characterization, making use of biological terms ("hormone secreted by medulla of adrenal gland"), and of its physico-chemical characterization, making use of physico-chemical terms ("aminohydroxyphenylproprionic acid—characterized by a structural formula for its molecules"). The case provides a specimen within the biology/physico-chemistry[6] example of branch reduction.

'Epinephrine' may be construed as a *biological* term, introduced historically by professional biologists via what Hempel calls its "biological characterization"; it may also be construed as a physico-chemical term, because (as I would put it), we now have reduced the concept to a complex of physico-chemical concepts: we now know its structural formula. Even before a detailed reduction was achieved, however, when it was reasonably believed that such a reduction would eventually succeed, one could have construed 'epinephrine' as a *physical* term—a term designating a *physical* substance (or physico-chemical substance) not yet fully understood physically.

To segregate a physico-chemical *vocabulary* from a biological *vocabulary*, to classify 'epinephrine' as a biological term—as if there were *linguistic* or *conceptual* difficulties involved in determining the *physical* constitution of the *physical* substance designated by the term—would surely present a misleading picture of the state of biological knowledge and research even at an early stage of inquiry. One may also well ask, what is meant by a *biological* (or physical or chemical) *characterization*? For Hempel, it is a characterization containing essential occurrences of *biological terms*. But what are biological terms? They are terms found in the specific vocabulary of the *biological sciences*. And what are they? Not the sciences dealing with biological *entities*, as Hempel clearly stresses (in the quotation given, above). What then? I suggest the only non-circular response is this: the notion of a biological science rests on a characterization in terms of certain historical features (see Chapter 7). But that is to say once again that *no philosophically significant vocabulary distinction* can be drawn here.

The present suggestion about interpreting the V_a/V_b distinction for branch reductions conveys an ontological implication of some interest.

I argued earlier (Chapter 7) that the reduction in the chemistry/physics case involved the replacement of *one domain* by *another domain*. In this chapter, I have maintained that there is but *one* domain, with physical and chemical poles, as it were. Have I contradicted myself? No. For a *domain replacement* is always the *result* of a *series* of *concept replacements*. To say that the "domain of chemistry" has been replaced by the "domain of physics" is simply to say that each of the *concepts* toward the "chemical end" of the spectrum has been *replaced* by a *construction* out of *concepts* toward the "physical end." Thus, one end—the chemical end—of what I am now viewing as a single domain has been collapsed or retracted into the other. (I ought perhaps here to remind the reader that these remarks bear on what I have referred to as the "theoretical attitude." From the practical point of view, the replacements involved may be viewed—and usually are—as *identifications*; see Chapter 4.)

One final example. Consider the relation between *astronomy* and *physics* as branches of science. I believe some further insights can be elicited from a consideration of the relation between these two. Traditionally, astronomy consists primarily of two distinguishable aspects: (1) the study of the *motions* of the heavenly bodies; and (2) the study of the *nature* (geological, physical, chemical) of those bodies. The first was incorporated long ago into (*reduced* to) classical mechanics; it is known as *celestial mechanics* and is simply considered now a branch of classical mechanics. It is in fact part of Newton's great unification of the "sub-lunary and super-lunary spheres." (We now use its relativistic version, since classical mechanics itself has been reduced to—replaced by—relativistic mechanics, a direct theory replacement reduction of the sort discussed in Chapter 5; for most *practical*, that is, computational, *purposes*, the classical theory remains quite adequate.) The second aspect has come of age more recently. It concerns the constitution and evolution of celestial bodies, explores such questions as: How are stars formed? What are they made of? How do they generate their energy? Why are they stable? What will become of them? How are planetary systems formed? This aspect of astronomy, known as *astrophysics*, may now be viewed as a branch of physics, or better, simply as the attempt to *apply* microphysical theory to a particular domain of objects: stars, planets, comets, galaxies, and various other exotic objects recently discovered (or hypothesized).

With respect to the issues at stake, the relation holding between (this second aspect of) *astronomy* and *physics* is quite similar to that between *chemistry* and *physics*. Yet, no one has ever discussed the former as an example of reduction (either of branch reduction or of theory reduction). The application of a V_a/V_b distinction (involving "astronomical terms" *vs.* "physical-terms") together with the attendant features of the standard analysis does not occur in the philosophic literature. It would not be inappropriate to treat our case as an instance of branch reduction, but it seems trimmer to view it merely as a case of applying a theory (microphysics) to one of its own subdomains (stellar structure and evolution). In this respect, it is hardly different from the chemistry/physics case. In both, the philosophic importance of a *vocabulary distinction* (a V_a/V_b distinction) is (at best) minimal. It would be idle to draw a distinction between a "physical" and an "astronomical" vocabulary for any but historical or perhaps practical purposes. The same is true in the chemistry/physics case (and in the biology/physico-chemistry case). (I should perhaps say once again that I have deliberately ignored such factors as instrumentation. Though it is surely true that the Mt. Palomar telescope is an *astronomical*, not a *microphysical*, instrument, that fact is philosophically uninteresting.)

I have been attempting to undermine the importance of a distinction featured in most current discussions of reduction. But in certain areas only. I maintain that the V_a/V_b distinction loses its significance when the '*A*' and '*B*' refer to selected *branches* of science. I also maintain that certain *branch* reductions may be so viewed that 'V_a' and 'V_b' merely designate portions of a continuous spectrum of cases. But not every example of reduction behaves thus. The V_a/V_b distinction continues to be valid and interesting in the context of certain *theory* reductions of the domain eliminating concept replacement type (discussed in the first four chapters), such as the thermodynamics/classical mechanics case and the (unsuccessful) electrodynamics/classical mechanics case. In those cases, one *can* formulate distinct vocabularies. There is a clear and precise characterization of *thermodynamic concepts* as opposed to classical *mechanical concepts*, or of *electrodynamic parameters*, etc. Nothing similar obtains in the chemistry/physics case. Everything depends on whether the distinctions of vocabulary or concept may plausibly be found in the subject matter of a given science and significantly bear on its being understood.

Why is it that a clear and significant vocabulary distinction exists in some cases (e.g., electrodynamics/mechanics) and not in others (e.g., chemistry/physics)? What underlies and accounts for the difference? It is not because the former involves *theory* reduction and the latter, *branch* reduction (although it is generally in branch reductions that the vocabulary distinction seems ill-founded—for reasons that will be apparent in a moment). It is rather a matter of the *degree of unification* of the "items" in question. For if both A and B (either as theories or entire branches of science) are *theoretically unified* and have an accepted set of postulates involving a small number of concepts from which all else (or nearly all else) flows, then A and B can be unambiguously characterized in a theoretical (not just an historical) manner. In turn, this means that specific vocabularies for A and B may be *recognized* (not merely stipulated) as the essential and characteristic terms occurring in a standard presentation of the theories or branches in question. If, however, A and B are each constituted not by a unified set of laws but rather by something closer to a "heap" of laws and generalizations formed for a variety of domains, then there is no clear *theoretical* way to characterize A and B; there would then be no clear reason for distinguishing the vocabularies of the theories or branches in question. A loose group of concepts and laws may eventually be organized into several *distinct* theories or branches of science as unification takes place. But, generally speaking, branches of science are less unified than specific theories.[7] That is why the V_a/V_b distinction seems more appropriate to the latter than the former.

But if this is so, then the V_a/V_b distinction should bear directly on *theory* reduction of the concept replacement variety. Hence, these present observations need not undermine the analysis of that type of reduction offered in the first four chapters of this account (where the distinction is taken for granted). It may be recalled that direct theory replacement reductions are homogeneous; but if so, there is at the very start no distinct V_b to discuss. So it is only in discussing *branch* reductions that one must be careful not to be misled by alleged general vocabulary distinctions. I believe my own, alternative reading of the vocabulary distinction satisfies this constraint in a less misleading way than the theory it replaces.

Finally, it should be remarked that a vocabulary distinction can be maintained as well for certain examples of reduction—drawn from logic, mathematics, metaphysics—introduced at the very start of this study

(Chapter 1). There, there are clear ways of distinguishing the sets of concepts under consideration. For instance, rational numbers are clearly distinguishable from irrationals—they do not shade into one another; there is no overlap. Similar remarks apply to the sense-datum / physical object distinction. On the other hand, there is no clear distinction between physical object terms and theoretical terms.[8] But this shows once again that distinctions of vocabulary of the kind here considered should be introduced and used only with the greatest care.

Notes

1. If university administrators can recognize that traditional departmental divisions no longer reflect the theoretical import they may once have had, philosophers ought to be capable of no less. I think here of SUNY Stony Brook's Department of Earth and Space Science, which now incorporates what were once viewed as quite distinct branches of science—geology, astronomy, oceanography.

2. It is tempting to label the penchant for drawing sharp vocabulary distinctions as another "dogma of empiricism"—or, perhaps, even a "dogma-schema."

3. See note 2 of Chapter 7.

4. See note 17 of Chapter 3.

5. Hempel, *op. cit.*, p. 181.

6. Notice that Hempel's use of the term 'physico-chemistry' may be used to sum up the point I am making. Think also of the import of such terms as 'biochemistry', 'biophysics'.

7. This is not always the case. Before Maxwell's great unification, electromagnetic theory was somewhat like chemistry before quantum mechanics. (Indeed, before that unification, one did not usually use the word 'theory'; one referred instead, e.g., to 'electrical science'.) Sometimes, as in the chemistry/physics case, one branch of science is unified only through its reduction to another; my suggestion that the terms of the reduced science may be construed as terms within the reducing science is most appropriate in such cases.

8. See, e.g., my paper "Theory and Observation," *British Journal for the Philosophy of Science*, May and August 1966.

Index